Simple Seismics

Simple Seismics

for the petroleum geologist,
the reservoir engineer,
the well-log analyst,
the processing technician,
and the man in the field

N. A. Anstey

International Human Resources Development Corporation • Boston

The author thanks the many institutions and companies who have allowed their material to be used in illustrations.

ISBN: 0-934634-37-8 Cloth
ISBN: 0-934634-43-2 Paper

Library of Congress Catalog Number: 82-80267

Printed in the United States of America

Contents

Introduction

This little book is different.

It is written, primarily, for geologists, reservoir engineers, and log analysts. Why? Because today's seismic method is more than a tool for reconnaissance exploration, for finding structures; it has become a tool for studying the discovered reservoir—its extent, its barriers, its variations of thickness, and its trends of porosity. Today, the geophysicist, the geologist, the engineer, and the log analyst can do great things together.

Because the book is not written primarily for geophysicists, it can skip much of the "mechanics" of the seismic method. The reader who reaches the last page (bless him !) will not be able to *practise* the seismic method, but he will understand how the seismic method can help to solve his problems.

It may also be of value to those practitioners of the seismic method—in the field or in the processing centre—who already know the mechanics of the method, but would like to take a broader view.

1

Seismics and Structure

Of all the geophysical methods working from the surface, there is only one which is widely accepted for studies of individual petroleum reservoirs—the seismic reflection method. We would dearly love to have something better, but alas...

Reduced to its essentials, the method is this: we make a bang, and we listen for echoes.

Just as if, on vacation at the Grand Canyon, we were trying to impress the kids by showing them how to measure the distance to the other side of the Grand Canyon: we might fire a pistol, time the echo, and use the known speed of sound in air to calculate the distance. We would be *echo ranging* with sound.

Seismic is sound in rocks. The seismic bang is a source of sound, the seismic wave spreading out from it is a simple sound wave in a solid, and the seismic reflection is an echo from a rock contact.

If then we can recognize the seismic echo, we can measure its two-way time, and form a judgement on the distance from the surface down to the reflector.

The problem of recognizing the reflection may not be a trivial one. In our Grand Canyon analogy, we would expect to hear the echo distinctly if the opposite wall of the canyon is sheer and plane, but we would expect a confused jumble of echoes if the opposite wall is very irregular. So we must accept that if the rocks under the surface are hopelessly contorted, the reflections from them may overlap and interfere so much as to be uninterpretable. There is a degree of geologic complexity for which the seismic method just throws up its hands and shrugs.

But, in the general case, let us say that we can recognize the reflection and measure its travel time. With that done, we can move to another position on the surface, repeat the observation, and deduce whether the reflector under that point is deeper or shallower than at the first point. And if we continue, making equally spaced observations along a line, we can build up a reflection profile showing the *structure* on the reflector.

Our basic scheme, then, is this:

● We fire a seismic bang at a first "shot-point" location SP 1 (Figure 1a).

● We record the reflections received at a geophone close to the source.*

● We repeat the observation at a series of equally spaced shot-point locations along a surface profile (Figure 1b).

● On a sheet of paper we draw a horizontal line to represent the line of profile, mark off the position of the shot-points 1, 2, 3,..., and then calibrate the vertical scale in seconds of two-way reflection time (Figure 1c).

● We display the seismic reflection received at SP 1, at the appropriate two-way time, on a vertical trace below SP 1.

● We repeat for each shot-point in turn.

The complete display is then a *seismic cross-section*—a representation, in units of time, of a slice through the earth along the line of profile.

On a seismic cross-section, the reflection indications go up

---

*In practice we offset the geophones from the source, but then we fiddle the results to look as though source and geophone were close together.

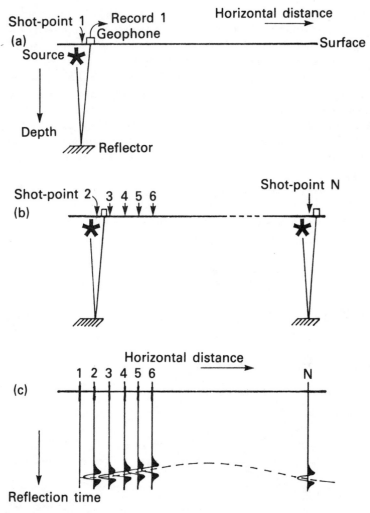

Figure 1 The basic concept: a reflection-time measurement at regular intervals along a line, and the representation of those measurements in the form of a seismic section.

and down with the structure of the reflector—we can see the structure. And we see it continuously, so that we are able to discover what happens between points of well control.

This, then, is the face value of a modern seismic section in an area of simple structure (Figure 2).

Confronted by such a section, the geophysicist tradition-
ally licks his coloured pencil, squints along the section, and
starts to "pick" the continuous marker reflections (as suggested by the dashed line in the figure). This part is fun.

In practice, he usually has a network or *grid* of seismic
lines; he then picks the same reflections on all the corresponding seismic sections (making sure, of course, that the reflections "tie" at the line intersections and around loops).
Somewhere in here, the fun starts to dissipate.

Then someone (nowadays, a digitizer) reads off the travel
time of each picked reflection at intervals of perhaps 5 shot-
points, and "posts" the values at the corresponding locations on
a base-map of the seismic grid (Figure 3a). These values can
then be *contoured*—by computer program or by hand—to yield
a *seismic structure map* (Figure 3b); if this is done by hand, the
fun returns.

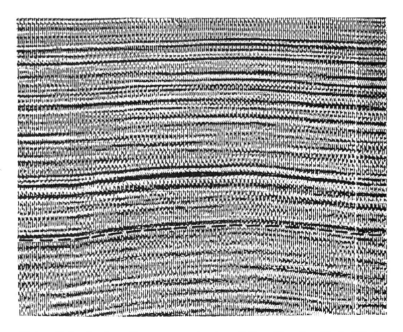

Figure 2 A seismic section across a simple anticline. (Courtesy United States
Geological Survey)

Well, you say, any ass could do that. And indeed, when seismic cross-sectional displays were first introduced (in the mid '50s), many geophysicists expressed fears that geologists and engineers and everybody would soon be picking seismic data—and what about job security? (They need not have worried; within a few years the digital revolution brought geophysicists an even better shelter—the mystique and gobbledegook of algorithms, and FFTs, and maximum entropy.)

So the picking of clear and continuous reflections, and the manipulation of the results into seismic structure maps, are really rather straightforward. However, everyone who even looks at a seismic section should be aware of three cautions—three respects in which the seismic section cannot be taken at its face value.

First Caution. When the seismic section is constructed (Figure 1), the reflection recorded at SP 1 is plotted *vertically below* SP 1. This is correct if the reflector is horizontal, because then the seismic signal travels down vertically, is reflected at right angles to the reflector, and travels back vertically. However, it is not correct if the reflector is dipping. Figure 4a shows that for a dipping reflector the reflection occurs in a zone which is *offset* from the vertical trace of the cross-section (Figure 4b).

This error, which increases with dip, is corrected in a mechanical fashion by the process of *migration.*[*] At its most advanced level this process is incomprehensible even to geophysicists, but at heart it is delightfully simple.

If all we have is the one observation at SP 1, and we have nothing else to tell us the dip, then all we know is that we have a reflection at a certain time. In that case, it could have come from anywhere along a surface (let us say a circle for the moment) representing that constant time. So in Figure 4c we actually put that reflection at all its possible sources around the circle. Then in Figure 4d we do the same for the trace from SP 2, SP 3...and so on. Our first thought is that this would produce a ghastly mess. But no—the circles reinforce in just one zone, and that zone is the true position of the reflector. In

---

[*]No connection with the migration of petroleum.

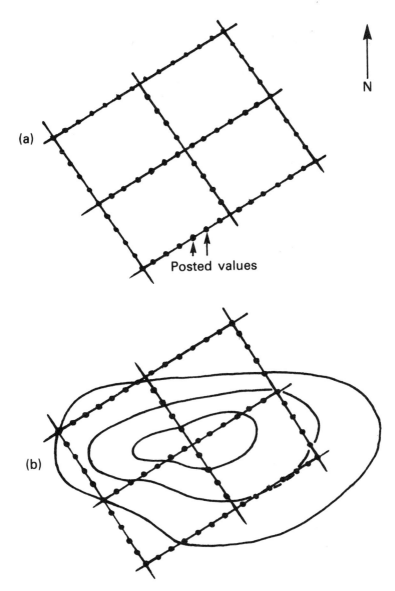

N

(a)

Posted values

(b)

Figure 3 The posting of reflection time values on a shot-point map, and the contouring of those values to yield a time structure map.

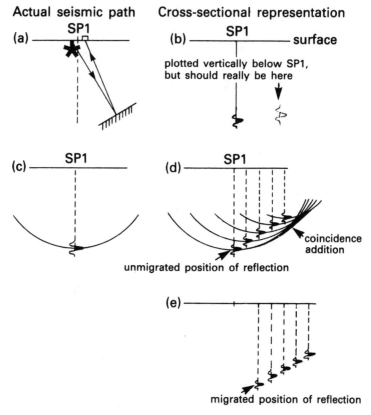

Figure 4 A simple view of the process of migration.

practice, we obtain a *migrated section* (Figure 4e) in which we see only the reflection moved to its correct position; the distracting "smiles" we would expect to see all over the section are usually seen only after the last reflection—when the section becomes dominated by noise.

The effect of the migration process is always to make the anticlinal structures less broad than they appear on an unmigrated section, and to make the synclines more broad (Figure 5). The axes do not move. We can, if we wish, perform the operations of picking, posting and contouring using mi-

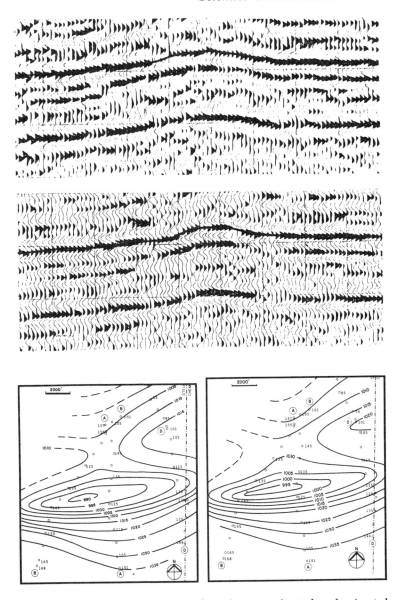

Figure 5 Unmigrated and migrated sections; unmigrated and migrated contour maps.

grated sections; then the shrinking of the area of the structural high becomes evident on the contour map.

Obviously, we must attend to these matters carefully before attempting to use seismic data in reserve estimates. Figure 6 illustrates the unmigrated and migrated versions of a seismic line across a structure formed by a local reversal of regional dip. The unmigrated section (Figure 6a) is sufficient to establish the position of the structural high, the position of the spill point, and the closure. However, the migrated section (Figure 6b) is necessary to establish the area of closure.

Second Caution. In all of this we have assumed that every reflection originates in the vertical plane through the seismic line—in the plane of the cross-section. We have made no provision for cross-dip, nor for local reflecting bodies (a salt dome, perhaps) off to the side of the line.

For reconnaissance surveys, and in areas of gentle geology, this does not matter very much; we lay out the important seismic lines as dip lines, and concentrate the interpretation

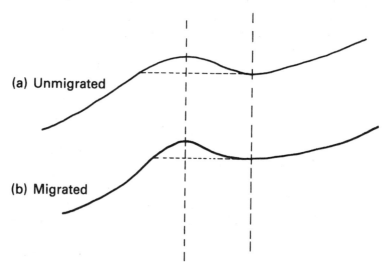

Figure 6 Unmigrated and migrated reflections. Note the broadening of the synclines and the tightening of the anticlines; the axes and the closure do not change.

on the migrated sections from those lines. To detail a roughly circular structure, for example, we like our lines to be radial.

But for more complex situations we are forced to a more comprehensive solution. This solution is based on the fact that the equal-time surface of Figure 4, from which we know that our reflection must come, is not actually a circle but a sphere.* We therefore perform a *three-dimensional migration* by distributing the reflection observed at each shot-point to all possible positions on a spherical surface centred at the shot-point. Then we do the same for all other shot-points, and see where the reflection indications reinforce. This may sound frightening. However, it is exactly analogous to the two-dimensional migration we discussed earlier, and in the same way the reinforcement shows the reflector in its true position—now in three-dimensional space.

To do this properly, we need to have much the same spacing of observation points across the line as we do along the line. In other words, we need a full 3-D survey in the field, typically consisting of a grid of lines spaced at 100 m or less. Such surveys are becoming commonplace for seismic studies of reservoirs at the detail stage. In complex situations, it is the only way.

Third Caution. The vertical scale of a normal seismic section is two-way reflection time. It is not depth. This is the big distinction over other forms of echo ranging; radar engineers, and officers of the watch, and bats, and dolphins, can all take echo time as effectively proportional to reflector distance—because the speed at which their radar or sonar pulses travel in their medium is substantially constant. For us, the speed of the seismic pulse in the real earth (we usually call it the velocity) is highly variable; there is a range of nearly 4 to 1 between the velocities of granite and newly deposited clay.

The classic hoax, in this context, arises in the search for reservoirs under salt. In Figure 7a we see what appears on the seismic section to be a very attractive situation—a broad anticline in a reservoir bed under a salt pillow. But the velocity in salt is high, while that in the overlying clays may be low. So

---

*A simplification for present purposes.

(a)

Apparent or
seismic time
section:

↓

t

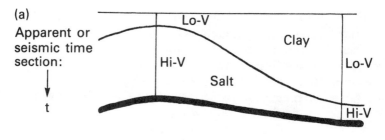

(b)

Real:

↓

z

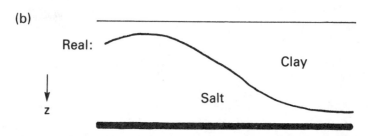

Figure 7 The classic velocity hoax: spurious structure introduced under a salt pillow.

the appearance of structure under the salt can be "velocity pull-up"—a complete hoax; the reduced reflection time under the crest is because of the high velocity, rather than because of shallower depth.

In the early days, these things were drilled. Some holes found the reservoir as flat as a pancake, and the geophysicist was rated a buffoon. Some found structure (though not as much as expected), and the geophysicist was half a hero. The key—both to avoid dry holes and to avoid missing a discovery—is to be able to *measure* seismic velocity.

The direct and obvious way is in a borehole—to measure the seismic time from the surface to a geophone at known depth in the hole. This is fine if we have suitably positioned boreholes (and we shall discuss the method in some detail later), but the real need is for a velocity measurement from the seismic observations themselves.

As usual in these matters, the details of the method are formidable, but the principle is delightfully simple. We shall be content with the principle.

As before, we take a measurement of seismic reflection time from a geophone close to the source (Figure 8a). Then we take another measurement with the source and geophone spaced apart; they are spaced symmetrically about the first measurement, so that the reflection occurs at the same place on the reflector— a "common depth-point" (Figure 8b). Of course, the reflection time is longer, because the seismic pulse

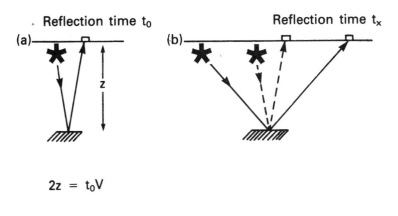

$$2z = t_0V$$

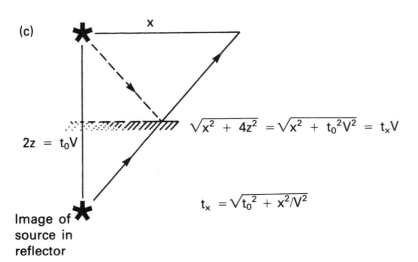

Figure 8 The concept of velocity analysis: the velocity is the extra distance divided by the extra time.

has travelled further. Then the principle of seismic velocity measurement is that the extra distance divided by the extra time gives the velocity.

At the uncomplicated level, it needs no more than Pythagoras. If we replace the source in Figure 8b by its image in the reflector, as in Figure 8c, the basic equation at the bottom slides out painlessly. This simplicity is at the heart of all seismic velocity analysis.

If we insert some values into the equation, it is easy to see that we should never again be hoaxed by the well-defined salt situation of Figure 7. Let us simplify the situation to that of Figure 9a, using the depths and velocities given there, and let us add another reflector C, within the clays, for convenience. The seismic time section appears as in Figure 9b with a major but spurious pull-up below the salt.

Then in Figure 9c we show the reservoir reflection as it is observed at the crest of the salt. At $X = 0$ we see the trace from the source and geophone close together; at $X = 1$ km we see the trace from the geophone spaced at that distance from the source. The zero-offset reflection time is 0.5 s, and the difference between the "near" trace and "far" trace is only 50 ms. That extra distance does not introduce much extra time, because the velocity is high.

In Figure 9d we see the corresponding picture off the salt. The intermediate reflection C (which has been chosen to have the same near-trace time of 0.5 s) now shows an increase of 175 ms on the far trace. Surely we could not miss a distinction as large as that. Even the reservoir reflection, at a near-trace time of 1 s, shows a far-trace increase of 100 ms.

We have to admit that it is a major weakness of the seismic method that it records reflection *time,* rather than reflector *depth.* In some situations we are definitely at risk because of this. Our hopes of eliminating the risk depend on studying the degree to which the reflecting time increases with the distance from source to geophone—on measurement of the "normal moveout".

Let us reconsider the reservoir of Figure 6. We have corrected the *area* of the petroleum accumulation, using migration, but we also need to know the *closure.* We already

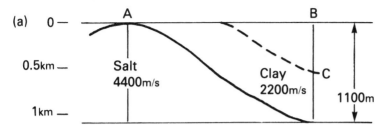

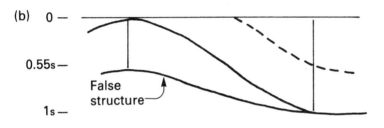

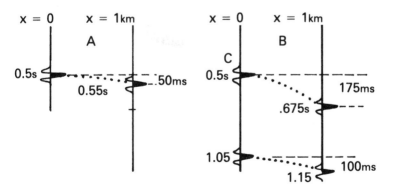

Figure 9 The solution to the hoax of Figure 7: comparison of long-trace times and short-trace times.

know it in time—in milliseconds—from the seismic section, but there is a risk that this is not a fair measure of the closure in metres or feet. There is even a risk that the closure does not exist—that the appearance of closure is solely a consequence of velocity variations.

So we conduct a careful velocity analysis. In practice we study not just two traces—near and far—for each reflection point, but a family of 12 or 24 or 48 traces at varying offset (a "common-depth-point gather;" Figure 10a). The gather on the crest of the structure (Figure 10b) probably shows more normal moveout than the gather at the spill point (Figure 10c); now we must convert these moveouts into velocities, using the equation of Figure 8, and compute the magnitude (and validity) of the closure in depth.

All of this sounds very neat, very scientific. But sometimes, when the wells are drilled, the geophysicist is still caught with his breeches at half-mast. What goes wrong?

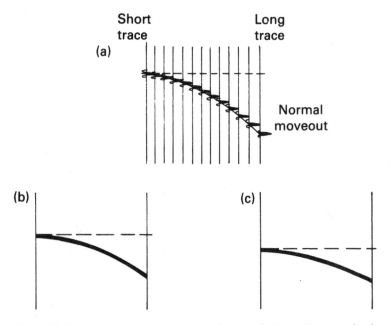

Figure 10 The development of Figure 9 for a multi-trace "common-depth-point gather."

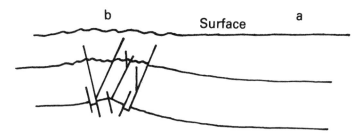

Figure 11 Unfortunately for velocity analysis, oil tends to occur in the complex geology at b, rather than in the layer-cake uniformity at a.

What goes wrong is the earth. In general, the techniques of velocity analysis, and of conversion from seismic time to real depth, work well in undisturbed flat-lying layer-cake geology. There they can be developed to great sophistication; the velocities of individual strata can be computed (even of quite thin ones, seismically speaking), and the process of migration can be refined to be exactly appropriate to the details of the velocity field. It is all very fancy indeed. Unfortunately, however, structural oilfields and undisturbed geology do not go together, and there's the rub.

Point to a on Figure 11, and ask a geophysicist about velocities, and you must be prepared to be smugged.* Point to b, and he will shuffle anxiously from one foot to the other. (Provided he is good; any geophysicist who would be complacent about his velocities at b is still awaiting his baptism of fire.) Ask him how many lengths of casing you should prepare for the completion, and you have lost a friend for life.

The problem is that the very circumstances which lead to tightly folded structural reservoirs and fault traps are the ones which introduce irregularity into seismic travel times. The reservoir reflector itself may be a rough erosional surface, or it may be broken into a number of small fault blocks. There may be extensive faults associated with the uplift; these bend, scatter, or otherwise distort the seismic paths. The compressional forces of the uplift may locally increase the seismic velocities as the rock particles are squeezed tightly together, or

---

*A new word meaning supercillied.

the fracturing and faulting may locally decrease the velocities as the particles are allowed to part. Erosional unconformities—and more bending of seismic paths—are likely at levels above the reservoir. And, often most serious of all, the surface expression of the uplift may result in irregular near-surface conditions which are difficult to correct precisely.

In such a situation, it is very dangerous to accept "velocity" values derived mechanically by application of the equation of Figure 8, and then to apply some sort of gross smoothing. The only hope is to study the individual common-depth-point gathers—a task requiring understanding, skill and great patience. This is the ultimate job security; if you're not a geophysicist, don't even fool with it.

At this stage, then, we have the following conclusions:

- In general, a seismic section showing clear and continuous reflection alignments can be taken at its face value—as a structural geologic cross-section.

- As such, seismic sections have a clear and obvious structural message which every geologist and engineer can safely be encouraged to see for himself. We should all be able to squint along seismic sections and grunt knowingly.

- There are three cautions—three respects in which the structural seismic section cannot be taken at its face value. These are: the need for migration in areas of significant dip, the risk that reflections originate out of the plane of section, and the fact that velocity variations in the real earth mean that reflection time cannot be taken as proportional to depth.

- Of these, the first two can be attended to—by intelligent placement of the lines and by the migration process, or at worst by a full three-dimensional survey. Although these processes have some dependence on velocities, we shall say here that this dependence is weak enough for us to accept that they restore the seismic time section to its face value—they are a complete solution to the first two cautions.

- In areas of horizontal beds and gentle structure, the velocities can be computed and the seismic time section can be converted to depth—to full face value. Then we have a complete solution to all three cautions. However, structural complexity—the very situation which most demands accurate velocities—may itself defeat the accurate measurement of velocity. To this there is no guaranteed solution.

- What do you want for $5,000 a kilometre—the earth?

There are two other matters we should discuss in this introductory session. The first is the question: What is it that *generates* a seismic reflection?

For present purposes, we can say that a reflection is generated at the interface between two rocks of *different hardness.* Thus although a reflection may be labelled with the name of a formation, it does not really come from the formation; it comes from the contact between that formation and the one above it. If the two formations are very different in hardness, the reflection is strong; if they are equal in hardness, the reflection is zero.

In the grossest sense, we can list rock types in a progression of hardness: clays, through sandstones, through limestones, to basement rocks. This would lead us to expect a fair reflection from the contact between clay and sandstone, or between sandstone and limestone; likewise we would expect a strong reflection from the contact between clay and limestone, or between sandstone and basement.

However, there are many factors which can materially change the effective hardness of a rock. Of these the most important is porosity (particularly if the pores contain even a small amount of gas). Thus a very porous liquid-saturated sandstone can show no contrast of hardness with a clay, and therefore generate no reflection. A reflection may reappear where the sandstone contains gas.

Of the other factors affecting hardness, those which make a rock more hard are compaction and cementation (and of course metamorphosis.) Microfracturing, overpressure and some forms of chemical change make a rock less hard.

Therefore a local change in the strength of a continuous reflection may mean a facies change, or a local development of porosity, or a change from liquid saturation to gas saturation, or a processing bust.

The geophysicist feels frustrated and cheated when a critically important geological contact does not show on his seismic section. However, this is a fact of life. In the North Sea's Auk field, for example, no detectable reflection is generated at the critical contact between the Zechstein carbonate reservoir and the overlying Upper Cretaceous Chalk. If there is no hardness contrast, there is no reflection.

The second additional matter we should discuss is the *appearance* of the seismic reflection. Squinting along the seismic section, we probably see it as one or more continuous black lines. If we look more closely, we see that on each trace it is a serpentine wiggle. This wiggle represents the up-and-down motion of the surface of the ground in response to the reflected seismic bang.

If, instead of displaying the wiggle on paper, we were to speed it up and play it through a loudspeaker, we would actually hear the bang...that is, the bang as modified by passage through the earth and the recording instruments, and followed by a train of echoes.

Intuitively, it is easy to accept that the reflection is more clearly located (better defined—*better resolved)* if the bang we hear is a short sharp click rather than a resonant ping or a dull rumbly thud. As every hi-fi buff knows, this is a matter of *bandwidth;* it is no good turning the bass and treble controls to anything other than "flat" if we wish to hear a sharp click as a sharp click.

Later, when we come to consider the ability of the seismic method to see small or thin reservoirs, we shall be much concerned with obtaining seismic reflections like short sharp clicks—with achieving good bandwidth.

2

Seismics and Stratigraphy

At this stage, we are comfortable with the operations of picking continuous seismic reflections, timing them at intervals along the line, posting the values on a base map, and contouring the result to obtain a *time structure map*. Further, we may do this for two or more reflections, and subtract one map from another to obtain an isopach map expressed in time—a *time interval map*. If the geology is such that we have confidence in our velocities, we may construct a *velocity map;* then (usually after an interpretation/smoothing stage) we may multiply this map by the former time maps to obtain *depth structure maps* and *isopachs*. But all of this—useful as it is—tells us next to nothing about the nature of the rocks. Let us see what we can do.

One approach is to attempt to measure from the seismic data the physical properties of the rocks. The physical properties which lend themselves to this, at the present stage of the established art, are velocity and hardness. Velocities are computed, as we have already discussed, from studies of normal moveout—the difference in time between near and far traces. Hardness is computed from the strength of the reflections; this is possible because, as we have noted previously, the strength of a reflection represents the contrast of hardness across an interface between two rocks. Usually the calculations are verified by setting up a *model* of the reservoir, with the expected velocities and hardnesses, and by checking that the seismic response of the model matches the observed data. This is the modelling approach.

Another approach is to use the observed reflection configurations —the patterns of reflections—to identify the conditions under which the rocks were deposited...even to establish the basin history. This is *seismic stratigraphy*.

These approaches are quite different, and in reservoir studies we need both. From the approach through physical properties and modelling we hope to learn the thickness and extent of the reservoir, possibly its porosity (for an assumed saturant), and possibly the position of any gas-liquid contact. From the approach through seismic stratigraphy we hope to learn the most likely lithology of the reservoir, the probable shape, the likely variation of porosity within the reservoir, the risks of permeability barriers—and where to look for another one like it.

We shall return to the approach through physical properties in Chapter 3. Here we explore the approach through seismic stratigraphy.

In our discussion of seismics for structure, we restricted ourselves (for simplicity) to well-marked and continuous reflections. Let us say we see such a reflection—strong, uniform and continuous—on line after line over the whole basin. What is the message? The first conclusion is that the reflecting sediments are almost certainly marine, laid down in water depth sufficient to give calm depositional conditions. This fact, and the strength of the reflection, probably suggest that the reflection comes from an interface between marine shale and limestone.

We see that the reasoning can be very simple—but very useful.

If the two formations are both massive, we obtain supporting evidence from their general appearance on the seismic section. For example, the two formations A and C in Figure 12, separated by reflector B, show a "grain" of smooth continuity within them—no strong reflectors, but the general aspect of

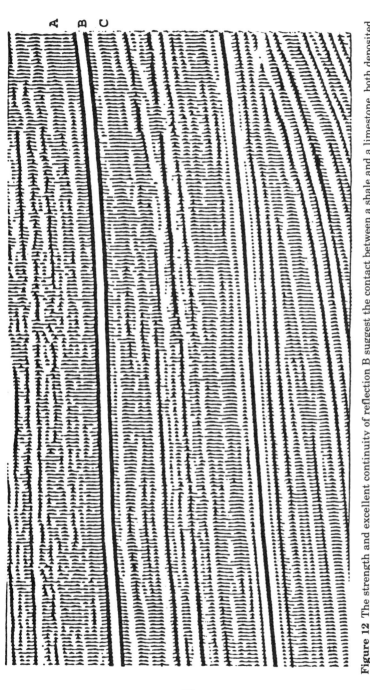

Figure 12 The strength and excellent continuity of reflection B suggest the contact between a shale and a limestone, both deposited under low-energy conditions. (Courtesy Prakla-Seismos Report)

23

calm deposition undisturbed by any locally variable geology. From this appearance we postulate that both formations are deposited under low-energy conditions, probably in fairly deep water. From the strength of the reflection we postulate that one is a shale and the other a carbonate. And from a velocity analysis we would hope to say which was which.

Our prime hope of a reservoir situation, on such a portion of seismic section, would be by the development of secondary porosity in the carbonate—by fracture, or by chemical change.

Let us suppose that we follow a marine shale, identified as above, towards the edge of the basin, and that there we find the situation shown in Figure 13. We see new reflections develop at the basin margin—new reflections which may be locally

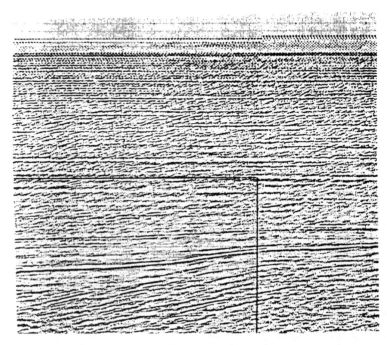

Figure 13 The onlap on the unconformity surface suggests marginal-marine sediments, deposited under higher-energy conditions. (From Leung; courtesy Amoco Oil Company)

strong but which do not persist for very far. Now our prime hope of a reservoir situation is in marginal-marine sandstones; we would be hoping for porous well-sorted sands deposited under conditions of high energy, and our first expectation would be that these sands would be oriented along depositional strike.

Perhaps we see, at another place on our seismic section, a substantial *absence* of continuous reflections—occasionally a local burst of strength, more often erratically weak reflections, but a general appearance without continuity (Figure 14).Then the depositional message is probably that these sediments are

Figure 14 The contrast between calm marine deposition (upper one-third) and non-marine deposition (lower one-third). The middle one-third shows some marine and some shallow-water deposition. (Courtesy Merlin Geophysical Company, Ltd.)

non-marine. Our hopes of reservoir situations turn to the sandstones and conglomerates of alluvial fans and the sandstones of braided-stream, channel-fill and other fluvial mechanisms; now we expect general orientation perpendicular to depositional strike.

These observations, we note, have nothing to do with seismic *structure*. They are concerned solely with the *appearance* of an interval on the seismic section. They may lead us to stratigraphic accumulations independent of structure, or they may need to be recognized in the presence of superimposed structure, or they may require interpretation in terms of reservoirs which have both structural and stratigraphic components.

Generally following Vail and his co-authors (1977), we may set up a formal relationship between the appearance of a portion of a seismic section (expressed in terms of the amplitude and continuity of its reflections) and the probable depositional conditions. At its most basic this relationship is:

● An interval within which there is a grain of low-amplitude reflections of good continuity suggests low-energy depositional conditions. The fact that the reflections are weak may mean that the continuity appears less good than it really is, because of noise; however, the grain of continuity should be visible. If the grain is conformable with underlying erosional topography (Figure 15), the deposition was from suspension.

● Reflections of constant *high amplitude* and *good continuity over large areas* (particularly when they represent the interface between the intervals just described) are likely to separate shales and carbonates deposited under low-energy conditions.

● Reflections of *widespread continuity* but highly *variable amplitude* (particularly if they exhibit angular relationships with reflections above or below) are likely to be erosional unconformities.

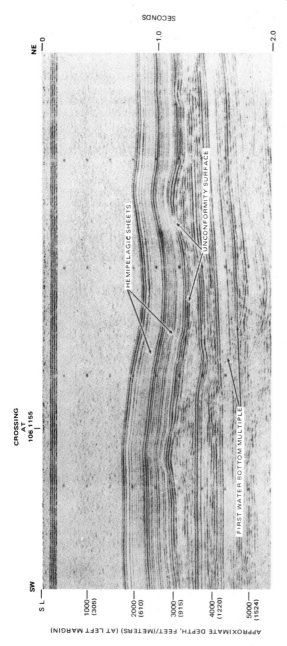

Figure 15 Conformable drape over a rough unconformity surface suggests fine-grained deposition from suspension. (From Sangree and Widmier; courtesy AAPG)

- Reflections of *high amplitude* but only *local continuity* in an onlap relation to an unconformity are likely to be marginal-marine sands or carbonates deposited during alternating high- and low-energy episodes. If the onlap relationship is actually seen at high amplitude, units of significant thickness are likely (Figure 16). If the onlap relationship is seen in the shale grain, but the high amplitudes are merely bulges on the unconformity reflection, then any high-energy sands which may be present are likely to be thin.

- Reflections of any amplitude having *local hummocky continuity* suggest deposition in shallow water (Figure 17).

- Reflections of variable amplitude and poor continuity suggest non-marine sediments (Figure 18). The continuity of a local reflection may appear absent on a cross-section,

Figure 16 A potentially prospective onlap relationship is seen at high amplitude about 1.6 seconds. (Courtesy Prakla-Seismos Report)

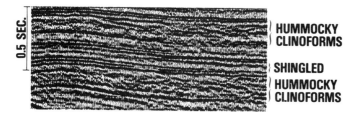

Figure 17 The "hummocky" signature of shallow-water sediments. (From Mitchum et al.; courtesy AAPG)

and its nature may be uninterpretable; study of its continuity in three dimensions may reveal a characteristic shape in plan, and so indicate the nature (for example, the fluvial nature) of the reflecting body. We shall return to this later.

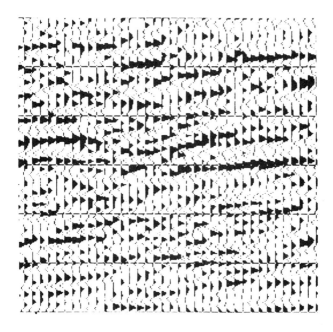

Figure 18 In the absence of tectonic complications, the poor continuity suggests non-marine sediments. (Courtesy Teledyne Exploration)

- Reflections of chaotic appearance *(variable amplitude, very poor continuity, variable dips)* suggest movement or slumping or mixed-energy fill.

- A substantial *absence* of reflections indicates basement, or salt, or the interior of some types of reef.

A seismic sequence is a stratigraphic interval bounded by unconformities or their correlative conformities. In Figure 19 we see three such sequences; they are separated by two sequence boundaries which are unconformities in some place and correlative conformities in others.

In this type of analysis, the picking operation on the seismic section is quite different from the one we used for structure. Instead of licking our coloured pencils and zestfully attacking the good continuous marker reflections (the fun bit), we gingerly start to mark the unconformity surfaces—the heavy lines in the figure. This is much more difficult, because unconformity reflections are by nature unsatisfactory; the materials in contact at the unconformity are always changing, and so the reflection—the hardness contrast—changes its strength and character also.

However, let us say that we do it. With some hesitation here and there, but we do it. Then we have split the section into

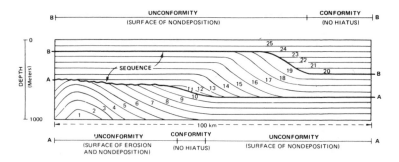

Figure 19 The tools of seismic stratigraphy: the external shape of a sequence, and the internal configuration of reflections within it. (From Mitchum et al.; courtesy AAPG)

sequences, and we note that each sequence has a *shape* in cross-section.

Next we mark the direction of the more-or-less-conformable reflection alignments *within* each sequence—the thin lines in the figure. Then for each sequence we have an *external shape* and an *internal reflection configuration*. Ideally we do this on a grid of seismic cross-sections, so that by making a fence diagram we can see the external shape in three dimensions. But for the moment we will stay with the two-dimensional display of the figure.

Now we are in a position to interpret the nature of each sequence.

The lowest sequence in the figure is a simple eroded uplift, with obvious possiblities for structural traps in units 1-4 and unconformity traps in units 5-10. Hopes of petroleum depend on the presence of a reservoir-type rock among these units, but the present analysis gives us little help in being sure of that. However, there is also hope of the development of porosity in weathered zones of the eroded uplift, and of course valley-fill sediments lying on the unconformity; the seismic picture would at least indicate the extent of the habitat of such possible reservoirs.

The middle sequence in the figure has a very different message. The internal configuration (and in this case the external shape also) indicates the upbuilding and outbuilding of a prograding shelf, during a period when the relative sea level was rising. The rock units to the right of the sequence must be predominantly low-energy marine shales; here the reflection continuity is likely to be good, but the amplitudes low. The rock units to the left are predominantly non-marine; the continuity is likely to be poor and the amplitudes variable. Between these zones is the clear sigmoid appearance of the prograding shelf; the rocks are shales, except for the flood-generated siltstone units which develop the sigmoid reflections.

Where are the reservoir habitats in this middle sequence? At the left, we have only the possibilities of braided-stream deposits, followed by point bars. Further from the left we have the possibility of river-mouth sand bodies and shoreline sand bodies. At the shelf edge we hope for shelf-edge carbonates—particularly reefs. Further to the right, the only significant possibility is deep-water turbidites.

The upper sequence is different again. Across the sequence boundary, between unit 19 and units 20-23, we have a clear onlap relation. One plausible inference is a rapid fall in relative sea level following the deposition of unit 19, followed in turn by a gradual rise accounting for the deposition of units 20-25. Transgressive shoreline sands are the likely reservoirs in this situation, on the rising onlap surface.

From this example, we can see the value of the seismic indications in fixing major rises and falls in relative sea level, and so in indicating the habitat and type of the reservoirs to be expected. Figure 20 illustrates characteristic highstand and lowstand situations, and the external shape and internal reflection configuration of the seismic sequences which they generate. By these means major rises and falls of relative sea level can be correlated all over a basin, or even at global scale.

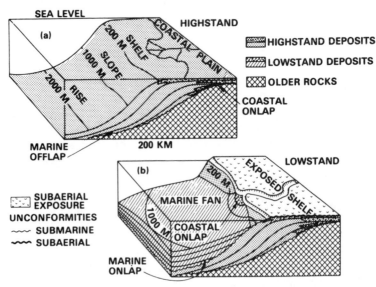

Figure 20 The seismic-stratigraphic signature of highstands and lowstands of sea level. (From Vail et al.; courtesy AAPG)

At this stage we can visualize the kind of seismic-stratigraphic analysis which would bring back the fun to this business—the fun we last enjoyed in a simple structural picking. Let us visualize a seismic section like Figure 21.

At (a) we see a major body whose interior is free of reflections. It could be salt, but the details of its upper surface suggest that it is crystalline basement (an interpretation which could be confirmed by gravity or magnetics). The importance of this is that we have a copious source of coarse-grained sedimentary materials.

At (b) we see another body without internal reflections—in fact its outline can be seen only as a break in the reflections each side. From the draping of sediments over it we know it is relatively incompactible; from its position we conclude it is a reef. From that, we interpret (c) as lime-rich fore-reef material and (d) as back-reef materials. At this stage we know nothing about the reservoir properties.

Above this level we need a sequence boundary, to signal the change to the prograding clastic sedimentation evident at (e) and (f). Within this prograding sequence we expect to find deep-water shales at (g), shales with some silt at (e) and (f), marginal-marine sands, silts and shales at (h), stacked mean-der-belt sands and shales at (i), stacked braided-stream de-

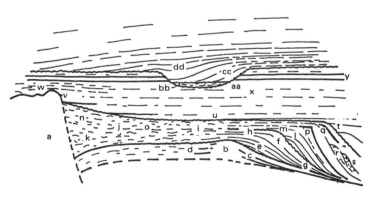

Figure 21 A hypothetical cross-section to illustrate how the depositional environment may be inferred for the external shape and internal reflection configuration of seismic sequences.

posits at (j), and stacked alluvial-fan sands and conglomerates at (k). The whole of the interval (i) (j) and (k) is likely to exhibit the scrappy appearance—variable amplitude and poor continuity—of non-marine deposits.

At (l) we see the evidence of a rapid fall in relative sea level, requiring a new sequence boundary. The fall is followed by a rise, leading to the onlap evident at (m); we now have the possibility of transgressive shoreline sands reworked from the previous sediments above (h) and (f).

The progradation continues, as does the rise in relative sea level. At (p) one of the river mouths bringing sediment to the sea is caught in the plane of section; instead of the previous *sigmoid* shape indicative of fairly low-energy conditions at the shelf edge, we see the *oblique* angularity generated at a delta as the river bulldozes its sediment load over the edge. In this locality, then, we have the increased attraction of prospects in delta-front sheet sands and delta-margin islands (both probably oriented generally along depositional strike) and in abandoned distributary channels and bar-finger sands (both probably oriented generally perpendicular).

At (q) we see a resumption of the sigmoid appearance, telling us that by this time the delta had wandered along the coast out of the plane of section. However, at (r) we see a submarine fan, which itself moved out of the plane of section and then came back at (s). In addition to the reservoir targets associated with the fan itself, we now have two more prospects; if we can find on another seismic line the then position of the delta at time (q), we can search for a sand beach between the two lines, and for the submarine canyon down which the fan sediments were transported.

At (t) we see the evidence of another fall of relative sea level, and we draw the appropriate sequence boundaries. Thereafter a major rise in relative sea level occurred; we see possibilities for onlapping shoreline sands at (t), (u) and (v). At (x) we see the characteristic signature of low-energy marine shale: weak amplitudes but a clear grain of continuity.

At (w) we see the possibility of sealed reservoirs in high zones of the weathered basement rock, or in coarse valley-fill deposits.

Thereafter, we will say, a limestone (y) is deposited over the area—thick, uniform and generally continuous, the kind of reflector which makes structural picking enjoyable.

Except that we can see at (aa) that it has been breached by a mighty channel. In the base of this channel, at (bb), is the potential for fluvial sand reservoirs; we see the chaotic appearance we would expect from high-energy fill.

Then we see evidence, at (cc), of sediment transport with a component from the right. At the level at which the limestone bed (y) is breached, these sediments are probably shale-prone; there is a chance that the present rightward dip of the limestone, coupled with a favorable bending of the channel, could provide a trap above (aa). (The bends in the channel, and its total configuration in plan, are very easily established on a grid of seismic lines; the channel is where the limestone reflection is not.)

The shale-prone nature of the sediments which fill the channel is evident from the shale grain, and at (dd) from the characteristic draping associated with the compactibility of the shale.

Figure 22 summarizes the major potential for petroleum reservoirs on this seismic section.

At (b) we can hope for a porous core or a porous crest in the reef. Further, there is likely to be a reef trend, possibly perpendicular to the plane of section.

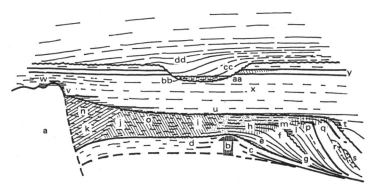

Figure 22 The potential reservoirs which may be identified from the seismic-stratigraphic analysis of Figure 21.

From the vicinity of (h) to the vicinity of (g), as suggested by the two hatched areas, we expect sand-prone shoreline deposits of regressive type; the conditions are those of a superabundance of sediments, a rising relative sea level, and low to medium energy. The alignment of these deposits is along depositional strike. The hatched zone is likely to contain a large number of long and thin sand bodies, some of which are separated by impermeable shale-breaks.

At (p), in the presence of a delta, we expect sand-prone delta-front deposits. Particularly attractive reservoirs may be formed by the reworking of such deposits in a high-energy environment, by the consequent formation of delta-margin islands, by the abandonment of sand-filled distributary channels, and by the formation of distributary-mouth bars and bar fingers. These possibilities are local to the delta, but the recognition of one delta on a prograding shelf suggests a search for others in a zone generally parallel to depositional strike.

Sand-prone shoreline deposits of transgressive type are likely to be found in the vicinity of (m), and from (t) through (u) to the vicinity of (v). These are all likely to be well sealed by marine clays. Again the alignment of the potential reservoirs is along depositional strike, and again they are likely to be long, thin bodies sometimes separated by impermeable shale-breaks. Above (q) there is the possibility that the transgression reworks and winnows (and later seals) the previous regressive shoreline sands; this is an attractive situation.

The general vicinity of (k) and (n) is likely to contain local reservoirs in the sands and conglomerates of alluvial fans. These bodies are very heterogeneous, and in the subsurface the detailed variations of reservoir properties are probably beyond the wit of man to unravel. The seismic character gives warning of this, but may not help to recognize individual reservoir bodies. Of course, the seismic obviousness of the mountain front (a) is itself an indication that alluvial fans are likely, and the directional trend of the fan reservoirs follows from that of the mountain front. Source rocks, however, are another problem.

Similar comments apply to the region of braided-stream deposition (j), (o) and (i). Although individual sand bodies exist—connected or unconnected—they are unlikely to be recognizable in the general fragmented appearance.

The region of the coastal plain (h-m) is the habitat of the meandering stream and the point-bar reservoir. Point bars are often of sufficient size and thickness to show directly on seismic sections. On any one seismic line, however, the best we can hope for is a short length of reflection; we must have a grid of closely spaced lines, and plot the extent of the reflection on a map, before we can see the characteristic shape of the point bars and their associated meanders. Because the appearance on a single line (fair local amplitudes, but only short segments of continuity) does not have any immediate appeal in a reservoir sense, it is important to recognize the *habitat* of the meander belt on the seismic section.

In the vicinity of (w), on the basement high, we have potential for small reservoirs in weathered basement or valley fills. The probable extent and orientation of both types of reservoir can be determined from seismic coverage, but little help can be expected in unravelling the expected variations of reservoir properties.

In the channel, we hope to find fluvial sandstone reservoirs of sufficient thickness to be evident seismically—as at (bb). Even if not, we would have some confidence that the wanderings of the channel could be located rather precisely from seismics, because of the breaching of the otherwise continuous limestone reflection (y). Of course, individual sandstone bodies, too small to see seismically but important as reservoirs, need not align along the channel but may wander across it.

Finally we have the potential for a clay-sealed trap against the wall of the channel, above (aa), if the limestone itself has reservoir properties.

We agreed earlier (and now we have demonstrated, on an artificial example) that one of the functions of seismic-stratigraphic analysis is to tell us something about the likely nature of the rocks. This it does by identifying local sediment sources, local sea level, local sediment-transport mechanisms, and local depositional environment. The important steps in such analysis are the identification of the sequence boundaries, the recognition of sequence shape, and the classification of the internal reflection appearance of each sequence. The analysis is not concerned with measuring the physical properties of the rocks—merely with defining the depositional conditions and making the allowable geological inferences.

Further, the analysis is affected by *structure* only to the extent that uplift and subsidence which occur during the deposition must be taken into account. For example, general basinal subsidence to the right of Figure 22 has both been caused by and allowed the greater thickness of sediments there. And the present leftward dip of the early alluvial-fan surfaces is also due to subsidence induced by weight of sediments; we see it as rightward dip in our mind's eye.

Where later structural activity is involved, it is usually still possible to recognize the sequence boundaries and the internal reflection configurations. Then it may be necessary to take out the subsequent structure (for example, by forcing a recognized shoreline back to the horizontal) to clarify the characteristic shape of a particular sequence.

Subsequent faulting, however, is more difficult. In effect, we have to try to restore the section to its unfaulted condition, with appropriate regard for the time, extent, and possible rejuvenation of the faults. Sometimes, it is tough.

Growth faults, of course, are an interpretation problem in their own right. However, they have their own habitat, and this can be recognized on seismic-stratigraphic considerations. They also have their own potential reservoirs, identified reasonably easily in seismic section.

So let us not dwell further on the problems, but carry forward the essential simplicity and appeal of seismic stratigraphy.

In so doing, we should recognize that stratigraphic seismics, like structural seismics, need not always show us the reservoir itself. If we do see a reflection complex of the right nature at the right place, well and good; even without that, it may still be very useful to identify a sand-prone environment, to establish its type and orientation, and to demonstrate the likelihood that it is sealed.

We said that seismic stratigraphy brings the fun back to interpretation. So the next few pages (Figures 23-28) should be one long chuckle.

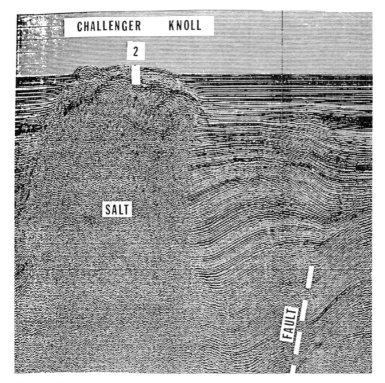

Figure 23 An easy one, for a start: the reflection-free interior of a salt zone. (Courtesy United States Geological Survey)

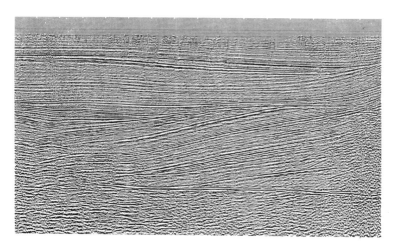

Figure 24 Exercise: trace the sequence boundaries. (Courtesy Prakla-Seismos Report)

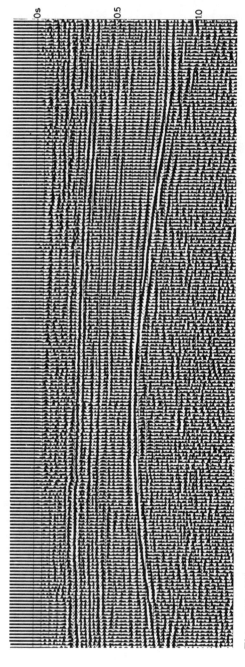

Figure 25 Where are the habitats of shoreline sands? (Courtesy Prakla-Seismos Report)

40

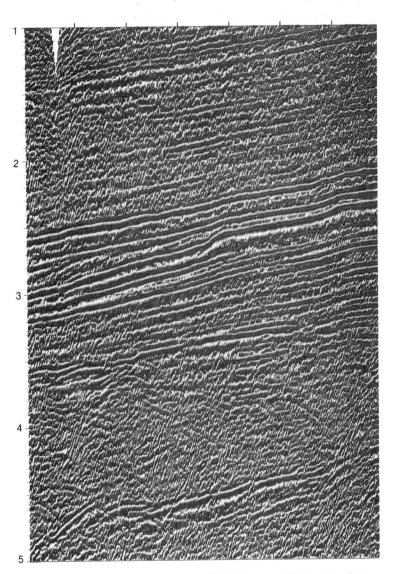

Figure 26 Where is the habitat of a reef? Of deltaic sands? (Courtesy Seiscom Delta Report)

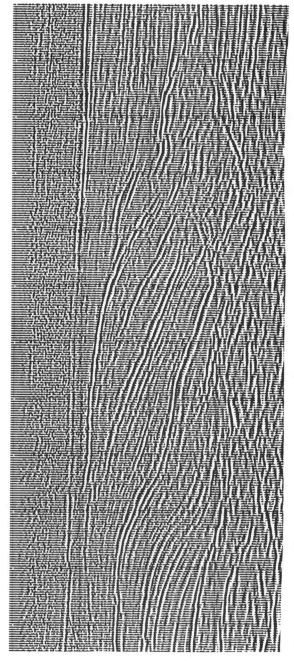

Figure 27 The grand-daddy of all prograding shelves. Slumping? A submarine fan? (Courtesy Prakla-Seismos Report)

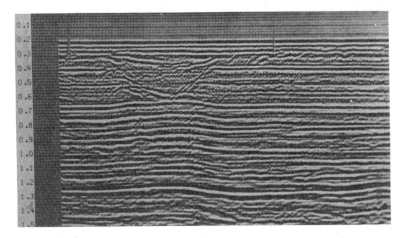

Figure 28 An erosional channel. In this unusual case, the sediments infilling the channel are of velocity *higher* than that of the breached sediments; note the velocity pull-up of the reflections below. The apparent structure may be entirely spurious.

3

The Message in the Seismic Trace

In Chapter 1, we discussed seismic data at large scale—that of the typical structure. In Chapter 2, we were at even larger scale—that of the whole sedimentary system. In the following chapters, we shall be at the scale of the individual reservoir. In this chapter, we look at the smallest scale—that of the single seismic trace. Our concern is to understand the connection between the rock properties and the seismic trace, and to gain some feeling for the risk in inferring one from the other.

We start with the assumption of layer-cake geology, for the present. We postulate the lithology shown at the left of Figure 29: a thick shale overlaying a massive limestone, with two sand units within the shale. The lower sand is completely water-saturated, while the upper contains both water and free gas.

Now let us construct the density log corresponding to this lithologic log. Working up from the bottom, we see first the significant difference in relative density between the limestone (let us say 2.8) and the shale (2.3). The porosity of the lower sand is such that, for water saturation, the density is the same as that of the shale. The porosity of the upper sand yields a slightly lower density where the sand is wet, and a much reduced density in the gas zone.

Next we consider velocity. The property of the rock on which velocity most depends is the compressibility; in general, rocks which are easily compressed have low velocities, and any circumstance which makes the rock more compressible (such as local porosity, or fracturing, or the presence of gas) reduces the velocity. We are not surprised, therefore, to find the variations shown under V-log in the figure: high velocity in the incompressible carbonate, a velocity a little higher than shale velocity in the wet sand, and a major depression of velocity in the gas sand.

In the last part of the Chapter 1, on structure, we introduced the other important seismic property of rocks—the hardness. Properly, this is the hardness in an acoustic sense; there is no guarantee that it conforms exactly to the geologist's scale of scratching hardness, though it probably does so approximately. Strictly we should call it acoustic impedance; the term need hold no fear for us, because the acoustic impedance is just the product of density and velocity (or, as

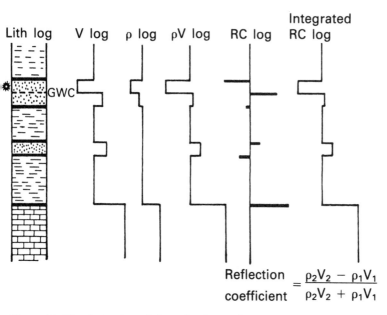

$$\text{Reflection coefficient} = \frac{\rho_2 V_2 - \rho_1 V_1}{\rho_2 V_2 + \rho_1 V_1}$$

Figure 29 The derivation of the reflection-coefficient log.

some people call it, the ρV-product). Thus, we get the ρV-log shown in the figure. It looks much the same as the V-log, except in the few cases where the velocity and density change in opposite directions. In Figure 29 this occurs in the upper wet sand, and is minor; in other rocks—particularly salt, gypsum, anhydrite, and coal—the distinction can be significant. However, at the first level of concern we often think of the ρV-log as being of substantially the same shape as the V-log.

Now we invoke again the fact that reflections are generated at the contact of two rocks of different hardness. We can draw the next log—showing the reflections generated at the rock contacts. The formula for computing the effective strength of the reflections is that for *reflection coefficient,* as shown—the difference of hardness across the contact, divided by the sum. Obviously, it is the numerator—the difference of hardness—which is important; a big difference of ρV makes a strong reflection.

And we note that the reflection is positive if the contact is from soft to hard rock, and negative if it is from hard to soft.

This is a good time to stress again that a reflection does not come from a formation, but from a contact. On the reflection-coefficient log (or RC-log) the value is zero within the formation—no matter whether the formation is a shale or a limestone.

But we are not limited to this if we do not like it. If, for example, we work down the RC-log making a running sum of all the values in a moving window, we obtain the integrated RC-log at the right of the figure; we have restored the appearance of the ρV-log, and now our readings indicate formations rather than contacts.

The RC-log of the figure is *the essential seismic trace.* The integrated RC-log may be regarded as *the essential seismic log* (or pseudo-log).

These two products are what we seek to obtain from the seismic method. It is important to realize that they tell us nothing, directly, about any property other than the hardness of the rocks. In particular they tell us nothing, directly, about lithology, or geologic age, or time-rock units. As we have seen in the last chapter, we can often make safe and useful inferences about these geologic features, because of their implications for hardness; in the final analysis, however, only the hardness counts.

To the degree that the seismic method is good enough to yield these two logs at the right of the figure, we can consider going one step further. In Chapter 1, we talked about measuring the velocity from the surface down to a reflector, using the difference of reflection time between the near trace and the far trace (Figure 30a). Then we could do the same on a deeper reflector (Figure 30b). Since each measurement gives us an effective velocity from the surface down to the appropriate reflector, we can see intuitively that we should be able to manipulate the values to give us the *interval velocity* between the two reflectors. The formula (for the simple case we are considering here) is obvious: the difference in depth divided by the difference in time.

If we were able to do this for each pair of reflectors in turn, we could construct the V-log of Figure 29. Then we could divide the integrated RC-log by the V-log, and so obtain the density log.

Further, if we could combine this information with knowledge of the lithology (for example, from nearby well control, or

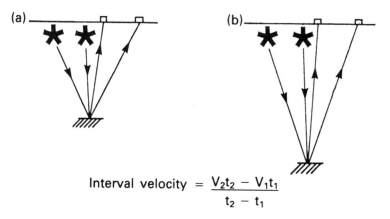

$$\text{Interval velocity} = \frac{V_2 t_2 - V_1 t_1}{t_2 - t_1}$$

Figure 30 The basic velocity measurement at two levels, and the computation of the interval velocity between them.

even from the seismic stratigraphy of the last chapter), we could compute the porosity.*

From this exercise, and possibly from the recognition of the reflection generated by the gas-water contact, we could be confident that we have gas in that zone; we could then compute the volume of gas, for a realistic range of water saturations.

So we have identified three clear objectives of the seismic method. The first, in Chapter 1, was the definition of structure. The second, in Chapter 2, was a contribution to the definition of stratigraphy. And now the third is clear: to make quantitative measurements of the properties of reservoir rocks, and so to allow estimation of reserves in place.

Those of us who are not geophysicists must be puzzled, sometimes, by the strange things that appear to be of concern to geophysicists. Anyone seeing a scholarly paper on maximum-entropy spectral analysis, for example, must be tempted to exclaim, "These guys have gone off the rails! What the hell does that have to do with real rocks?"

Be comforted. Hidden behind all the mish-mash is a highly directed effort to remove all the "ifs" of the above sequence of rock-property measurements—in particular, to obtain seismic records which look like the RC-logs of Figure 29, and to make velocity measurements over the thickness intervals of a reservoir. These efforts are critical to the advancement of rock-property analysis from seismics.

We see, then, the best that the classical seismic method can hope to provide in this context: reservoir extent, reservoir thickness, reservoir porosity, and the position of the gas-liquid contact. Oil-water contacts are not generally visible, and we see no hope of a seismic measure of permeability.

With these objectives and hopes established, we must spend a little time understanding the barriers to their fulfillment and the errors to which they are subject.

The first problem is "noise." Obviously, if a movement occurs at the surface of the ground, and that movement is *not*

---

*Using the equation $\phi = (\rho_g - \rho)/(\rho_g - \rho_f)$, where ρ_g is the grain density of the rock, ρ_f is the density of the saturating fluid, and ρ is the bulk density as measured.

caused by a reflection from within the earth, then we are at risk; the movement is picked up by the system, and used to suggest a variation in rock properties deep below. We need to be able to trust every last little wiggle of the seismic trace as being genuinely caused by reflections.

Nowadays, in many areas and for many purposes, we are there. The massive field techniques of today—in which 24 or 48 or 96 records are taken for every one strictly necessary— often give us that confidence in every wiggle. For most structural purposes, and for broad analysis of seismic stratig- raphy, the problem is beaten. For the calculations of reservoir properties, as we shall see later, it is not.

The next problem is multiple reflections—reflections which have bounced two or three or more times, and which therefore arrive at the surface at a time which we must not accept at face value. Again, in most areas and for many purposes, we are there; the field techniques, coupled with sophisticated processing, have virtually beaten the problem. Usually, we can take it that the reflections we see on the seismic section are single-bounce primary reflections.

The next problem is the near-surface. The problem lies in correcting the seismic reflection time for the ups and downs of the surface, and for local variations of near-surface velocity caused by the water-table, by alluvial and glacial fill, and by weathering. Present solutions of this problem are usually sufficient to provide us with good seismic cross-sections, on which we can trust the indications of large structure and of seismic stratigraphy. We can be less sure with structures of low relief, and we are positively nervous if these occur in the same place as some obvious near-surface anomaly.

But it is when we come to the measurement of velocities (either for time-to-depth conversion, or for calculation of reservoir properties) that we are most worried about the near- surface. This is because then we are taking the difference of two times to give us the velocity. In Figure 31a we recall the basic method of measuring velocity, from the difference be- tween the far-trace time and the near-trace time. In Figure 31b we illustrate that the calculated velocity could be much too low if there is a local low-velocity anomaly (for example, a dry

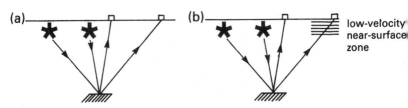

Figure 31 The sensitivity of velocity analysis to local delays in the near-surface zone.

alluvial fill) under the far geophone. The same fill under the *near* geophone makes the velocity *too high*.

The sensitivity of velocity analysis to local anomalies in the near-surface (or indeed local anomalies anywhere above the reflector of interest) remains a major current problem of seismic work. This is one of the present limitations to the seismic calculation of reservoir properties.

Finally, among the problems, we have the problem of the seismic pulse.

Thus far, we have said little about the seismic signal except that it is some sort of bang. Let us take the classic seismic source—a dynamite explosion—and ask our intuitions what we might expect to get from it. First, we would expect that the explosion pushes the surrounding earth outwards; the earth goes outwards and stays outward, leaving a hole blasted by the explosion (Figure 32 at (a)). Further away, however, we would expect that the earth is pulsed outwards but that it then settles back (b); in particular, by the time the bang has travelled down to the reflector, been positively reflected, and come back up to the surface, we would expect that the surface of the earth would take a little jump upwards, but then settle back again.

Of course, after all that travel it would be only a little jump. But suppose it were just big enough to be felt with the flat of the hand—how long would we guess the jump up and settling back might take?

Looked at it this way, the right answer—something between a tenth and a twentieth of a second—seems very reasonable. For something of the scale and nature of the surface of the ground, we really would not expect it to be much less.

As we try to sense how it might feel on the hand, we notice that we cannot move the hand up and down again, so quickly, without overshooting. And so it is with the earth; the minimum final response (c) of the surface, for a simple outwards movement from the explosion, is up-down-overshoot-back. All in a tenth or a twentieth of a second (d).

Normal geophones are sensitive not to the up-down displacement of the surface, but to the rate-of-change of this motion. So the geophone signal (e), in response to a simple reflected bang, typically consists of one-and-a-half cycles, having a total duration of the order of one-tenth to one-twentieth of a second.

In fact, there are several agencies which contribute to this response; however, the simple picture will suffice for all our present purposes. From one raw bang we get a pulse of one-and-a-half cycles (probably with some more small ones caused by the other agencies) and of about 50-100 ms duration.

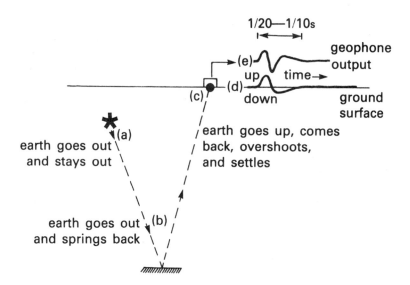

Figure 32 The evolution of the wiggly nature of the seismic pulse.

The full horror of this comes home when we recall that seismic velocities are typically several metres per millisecond. The bang, as it comes up to the surface, is stretched across perhaps a hundred metres. We visualize a typical well log (Figure 33, left) as the pulse goes past; there is the full horror. The seismic pulse in the earth is many times longer than the thickness of most of our reservoirs. A blunt instrument indeed. Each rock contact on the log reflects the whole pulse, to a degree depending on the contrast of hardness across the contact. No surprise, then, that a seismic reflection trace is a complicated thing, representing as it does the superposition of a very large number of overlapping reflected pulses from a profusion of geological contacts. To understand the seismic trace, we need the *synthetic seismogram*.

In Figure 34a and 34b we reproduce the lithologic log and the RC-log of Figure 29. We have six reflectors, represented in time, strength and sign by the spikes of the RC-log. For each of these, Figure 34c-34h, we draw a reflection pulse—starting at the appropriate time, scaled to the appropriate strength, and erect or inverted according to the appropriate sign. Of course, they overlap. Then we just add them together to obtain the synthetic seismogram shown in Figure 34i. This is the seismic response to the given lithologic log, for the stated form of the seismic pulse.

The trace of Figure 34j is a repeat of the RC-log, for easy comparison with the synthetic seismogram. The relation between the two is obscure, to say the least. For one thing, the obvious part of each pulse is delayed, relative to the expected time on the RC-log. Further, it is not clear what part of the pulse we should use for our judgements of the time and strength and sign of the relections.

Fortunately, the processing wizards are rather good with this sort of problem. "What we'll do," they say, "is to change the shape of your pulse, in the computer, so that you obtain the record you would have had if the bang had generated this new pulse shape. Oh, and while we're at it, we'll fix that time delay, and make the pulse symmetrical about the correct time (so you know where to pick), and make the centre of symmetry black

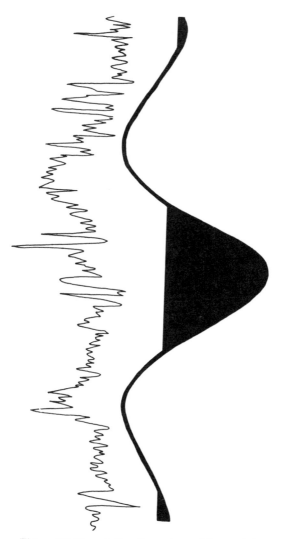

Figure 33 The relative dimensions of the earth layering (as shown by the well log, left) and the effective seismic pulse. Ouch.

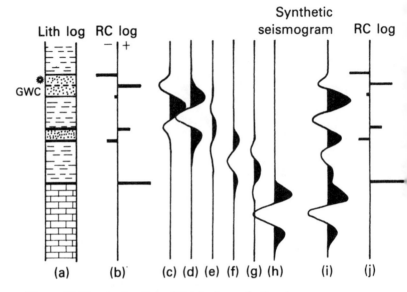

Figure 34 The construction of the basic synthetic seismogram.

for a positive reflector and white for a negative reflector, and ...and so on. It's part of what we call *wavelet processing*. Just leave it to us."

So we give our seismic traces to the processors, and they produce Figure 35 as their modified version of Figure 34. Each pulse is now symmetrical and centred on the correct time. The synthetic seismogram—the result of adding the six pulses—now looks much more like the RC-log; we have a black peak more-or-less corresponding to each positive contact, and a white trough more-or-less corresponding to each negative contact. In particular, we are delighted to see the clear black peak of the gas-water contact.

Unfortunately, we have one black peak and a couple of white troughs which do *not* correspond to a contact, which do not have any useful or geological meaning. By inspection, we can see that these hoaxes are consequences of the wiggly nature of the seismic pulse, which is in turn partly due to the

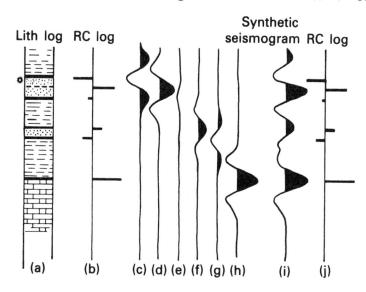

Figure 35 The seismogram is more directly interpretable, in terms of rock contacts, after wavelet processing.

fact that the earth settles back after we have banged it. We can guess that this effect will be difficult to remove; it is. We shall talk about it a little in Chapter 7.

For the present, we turn to Figure 36, in which the first two traces merely repeat the last two of Figure 35. The fourth trace is the integrated RC-log of Figure 29; this is the log which most clearly represents layers rather than contacts. Then if we do the same integrating trick (let us think of it, for present purposes, as a simple running sum) on the synthetic trace at the left, we obtain the third trace—the first step towards the synthetic seismic log or synthetic pseudo-log. This is our attempt to make the seismic trace represent layers.

It is not too bad. The soft layer (a) clearly deflects to the left; the hard layers (b), (c) and (d) deflect to the right. But still we have the problem of those outriders each side of the main indication; each side of the correct deflection we have smaller opposite deflections which are quite spurious.

Synthetic   Synthetic   Integrated
seismogram RC log pseudo log RC log

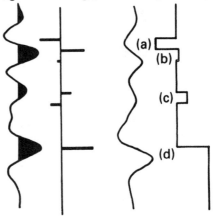

Figure 36 An approximate view of the construction of the pseudo-log (or seismic log, or acoustic impedance log).

So we must be aware of this fundamental problem whenever we look at a seismic trace—whether the trace is in conventional form or pseudo-log form. However, the problem should not cloud the basic conclusion of this chapter: that *the synthetic seismogram gives us the seismic trace we can expect to obtain from a defined series of layers in the earth,* for a given form of seismic pulse.

The synthetic trace is correct only if the pulse is correct. Let us spend a moment to see what difference it makes.

Figure 37 repeats the operations of Figure 35, but for a rather sharper pulse. Now we see a clearer and better correlation between the contacts on the RC-log and the peaks and troughs of the synthetic trace. The hoaxes are still there, but at least part of the message is good.

We note, too, that the time thickness of the gas column in the upper sand just coincides with the half-period of the pulse, and that this produces constructive interference (or *tuning*) between the top-sand and gas-contact reflections.

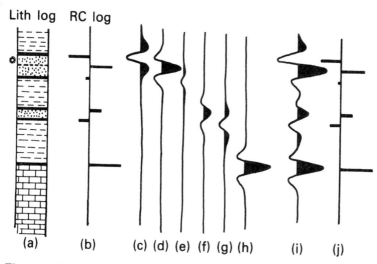

Figure 37 The construction of the basic synthetic seismogram, for a rather sharp pulse.

Let us take it further. In Figure 38 we display the synthetic traces for a range of pulse shapes; the synthetics of Figure 38d and 38e are those we have seen previously, and now we add one with a very poor pulse (Figure 38c) and one with a very good pulse (Figure 38c). We might think of the sequence of seismic signals which generates these four traces as: a dull rumble, a thud, a bang, and a sharp crack.

And we see how important the pulse is. The dull rumble does give us the top-lime contact, but it obscures the gas sand and effectively loses the water sand. The thud gives us some indication of everything, but it makes both sands appear thicker than they are. The bang does quite well; it gets the thicknesses right, but it exaggerates the gas-water contact by the tuning effect. If we really want to see what is there, and in particular if we want to make measurements of the reservoir properties, we must ask for the sharp crack.

"Alright," say the geologists and engineers, "but that's enough of that. What's the *use* of all this synthetic stuff?"

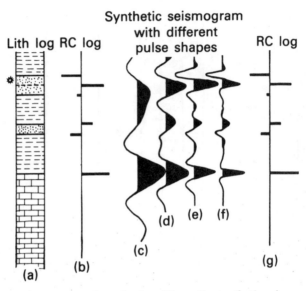

Figure 38 The effect of the pulse breadth on the synthetic seismogram.

The three main inputs to a synthetic seismogram are the V-log, the ρ-log, and the shape of the seismic pulse. If we have well control, we have the first in the form of the sonic log, and the second in the form of the density log. The pulse we can regard as the variable (as we did in Figure 38), or if we have a seismic line in the area we can extract it from the seismic data (using some of the processors' magic programs). So in many reservoir situations we can synthesize the expected seismic response of the reservoir zone. Make no mistake about it—the usefulness of this is real.

- First and foremost, it allows us to *understand* why a certain piece of earth gives a certain seismic trace.

- Next, it allows us to identify the reflections positively. We must always remember that, because seismic velocities vary from area to area, geophysicists working in a new area cannot tell just by looking at the reflections what geological contacts they represent; we may know the expected depths, but that does not give us the expected times.

● It tells us *whether or not we shall see the reservoir*. If not, it tells us why not. If the reason is that the contrast of 1ardness between the reservoir and its bounding rocks is 1adequate, then we must accept that there is nothing we c n do to study the details of the reservoir seismically; we sł 'uld spend the money drilling holes. If the reason is that the reservoir is too thin, or too close to some very strong reflector which swamps it, then we must address the problem of the seismic pulse; we must try to change it from a lolloping rumbling thud into a short sharp crack. In Chapter 7, we shall see that this can be done, but that it costs money. So we have to study the relative cost-effectiveness of this approach and the drilling of holes.

● If we do see something on our seismic sections at reservoir level, the synthetic seismogram tells us what part of the reservoir it represents. In particular, the synthetic tells us where to look for the all-important gas-liquid contact.

● A group of synthetics tells us how the seismic response changes as the reservoir changes. For example, let us ask how the reservoir response of Figure 37 changes with the height of the gas column. The "tuned" synthetic of Figure 39 is the same as the reservoir portion of Figure 37; we remember how it gave a correct indication of the time thickness of the gas zone, and how the tuning effect exaggerated the amplitudes. Then the other traces synthesize the changing seismic response to the height of the gas column, with less pay to the left and more to the right. If this is what we see on a seismic line through the well, the message is obvious; we know the degree of thickening to the right, and we can project the limit of the gas to the left. In the same way, we can predict the seismic response to an increasing thickness of channel sand (or to the shaling-out of a stratigraphic reservoir, or to the

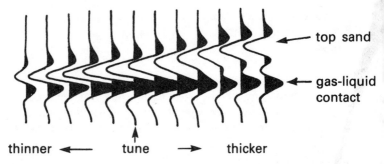

Figure 39 The synthetic seismogram used to explore the details of the reservoir (in this case, the thickness each side of the "tuning" thickness).

insinuation of a carbonate bank, or to a 5 m fault) and see whether this change can be found on the real seismic section.

● Finally, the synthetic is the essence of the *modelling* step, by which we check the validity of any detailed interpretation of seismic data; from the final interpretation we synthesize the seismic response, and see whether it checks with the data.

From all of these emerges a regimen for the seismic study of a reservoir. We shall set out and illustrate this regimen in the next chapter.

4

In the Mind of the Interpreter

As we made clear in the introductory material, the object of this little book is not to turn specialists of one discipline into specialists (or even practitioners) of another. Specifically, this part of the book is not trying to turn geologists, engineers and log analysts into seismic interpreters. We seek to convey only that information which will allow these other specialists to understand what the seismic interpreter does, the strengths and weaknesses of it, and the degree to which the weaknesses can be removed by more exhaustive (and costly) developments. From this we hope that many of the problems typically encountered in exploration departments (specifically, the oft-recurring one: "Well, what should we do—shoot more seismic, or drill another hole?)" can be tackled cooperatively and cost-effectively by all the specialists involved.

So let us spend a little time looking over the shoulder of the seismic interpreter as he practices his craft.

It is Monday, the first of the month. A bundle of sections drops on his desk. They show evidence of very hurried printing and packing; they are shuffled, out of order, different ways up, folded where they should have been rolled, or folded face inwards where they should have been outwards. This is because the contractor needed to get the work out before the end of last month, so that he could invoice, and because his staff knew that if they did not get it done by Friday afternoon, they would have to work the weekend. However, none of this fazes our interpreter; he is just thankful that his sections were not sent to Mogul Oil by mistake.

By noon, everything is back in order, and the work can start. Probably the first thing he does is to work carefully through the sections to see that they *look* right. During this operation, the interpreter notices that one section does not look right; the direction of the line has been labelled northwest where it should have been southeast. (He is always very hot on this kind of error, because his company once drilled a dry hole on a mis-labelled line.) More generally, however, he is visually making an assessment of the broad degree of confidence he can place in the data—do they look seismically trustworthy and geologically reasonable?

Obviously this is a subjective appraisal, based on his experience rather than on any communicable skill, so we will come back when he has finished. At that time, he is starting to scan for structure.

He starts, as we did earlier, by accepting the seismic section at face value—as a geologic section. He squints along the good continuous reflections, looking for closure (relative to the horizontal timing lines). Having found closure on one line he searches for the corresponding feature on other lines, and perhaps arranges the dip sections to align the structural axis (Figure 40). Then he looks for four-way closure—by searching for closure on the strike sections, or by comparing reflection times on several dip sections. He may find, of course, that there is the four-way closure of a closed anticline on one or more reflections at depth, but that shallower in the section there is only the three-way closure of a plunging anticlinal nose. Then he must search for evidence of updip fault closure on the nose.

After a rough appraisal of this type, the interpreter starts the picking operation, colouring an appropriate part of the reflection waveform. When this is done on all sections, and the consistency of the picking is verified at all the line intersections, the picked times can be posted and contoured (as described in connection with Figure 3.)

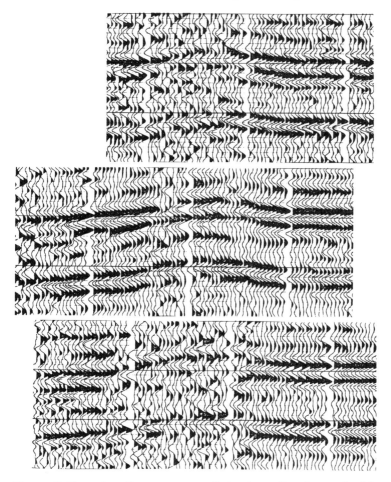

Figure 40 Three short lines across a small structure, aligned approximately at the anticlinal axis.

At this stage the interpreter might take a first counsel with his colleagues in the exploration department: Is this structure (at its face value) large enough to be interesting?

If it is, then the interpreter must start to consider the deviations from face value—he must acknowledge that a standard seismic section is *not* a geologic section. In so doing, he will probably pose several questions to himself—questions which the others of us (if we wish to appear erudite in an exploration meeting, or if we wish to get the measure of our geophysicist) might well pose to him too.

Is there any chance, he will ask, that the reflection I am picking is a multiple? This is material, because if a shallow non-prospective reflector has slight closure, its multiple reflection (bouncing off the underside of the surface) has more closure; as suggested in Figure 41, there is the potential for an attractive-looking hoax. The risk is real. The protection against it lies in having an interpreter who understands what has been done (and the human judgements which have been made) during the computer processing of the data, and who takes the time to check that those judgements were correct.

Is there any chance, he will ask, that this apparent structure is merely a consequence of the near-surface—that the reflector itself is flat, but that improper correction for a hill (or a valley, or a swamp) shortens the travel times over the apparent crest or lengthens them over the apparent flanks? Figure 42 illustrates typical situations where this risk arises.

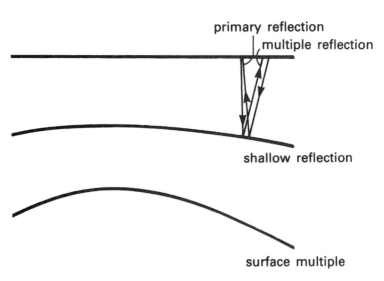

Figure 41 The multiple of a shallow anticline may appear as an even more attractive (but quite spurious) anticline.

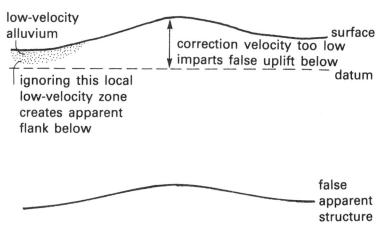

Figure 42 The risk of drilling false structures introduced by the near-surface.

As a first protection, our interpreter insists that the processor add at the top of each section the elevation profile of the surface (or the water depth if the prospect is offshore); he may also call for indications of the position of rivers, moraines, outcrops, etc. Then his initial check is to see whether the apparent structure is in correlation (or possibly anti-correlation) with the surface. Of course, a correlation between the subsurface and the surface is often geologically genuine; but now the interpreter must go through the exercise of proving that it is so in this case. The proof may be simple (if the apparent structural closure is far too large to be caused by surface effects) or it may be very involved. Indeed, it may not be possible without conducting an independent survey of the near-surface over the area of interest. But it is essential that the interpreter satisfy himself one way or the other; we are not yet immune from the risk of drilling false structures caused by the near-surface.

Is there any chance, he will then ask, that this apparent structure is caused by lateral variation of the average velocity down to the reflector? We discussed one situation where this can occur—a salt swell—in Chapter 1; Figure 43 reproduces the essentials of a previous illustration. The interpreter therefore scans the section for the reflection-free appearance

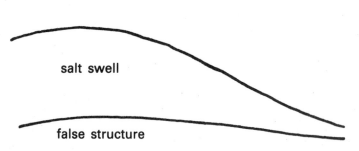

Figure 43 A reminder of the salt-swell hoax.

which (as we noted in Chapter 2) characterizes the interior of a salt mass. If he finds such indications, he then checks with gravity/magnetics to verify that it is indeed salt. Finally he computes, using the measured or known velocities, the degree of falsification of the closure which is due to the salt.

Or, the lateral variation of velocity may be due to fracturing; he may be confronted with the situation of a previous figure here paraphrased as Figure 44. In this case, though, the risk is in the opposite direction; usually the fracturing occurs over the uplift, and so the consequent depression of velocity causes the closure to appear less than it really is. Anyway, our interpreter must go through the exercise of trying to get reliable velocities from the seismic data, and then of correcting the apparent structure from time to depth in order to see the real closure. As we noted earlier, this can be very difficult (and sometimes impossible) because the consequences of structural uplift often defeat good velocity measurement. We are not yet immune from the risk of drilling false structures caused by lateral velocity variations; probably we never shall be.

Is there any chance, our interpreter now asks, that the structure I am seeing originates out of the vertical plane of section? The first test he applies here is geological reasonableness; if one reflection passes straight through another, or if

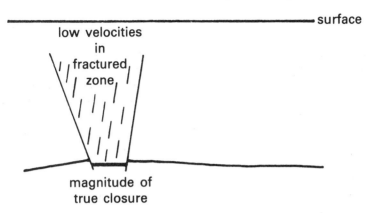

Figure 44 The risk of falsification introduced by fracturing and its attendant low velocities.

an anticline appears out of geological context, he is on guard. (In a particularly abstruse case, the resolution of the problem may even force him to make a major sacrifice for science, and to take a geologist to lunch; the ploy is to sketch the situation on a napkin over dessert, and to inquire artlessly,"Do you think that can happen, in real geology?") Of course, the problem occurs only on the outside lines of a survey, or if the lines are far enough apart to allow some side-swiping feature to lurk unknown between the lines; in a survey to delineate a drilling prospect this side-swipe problem is unlikely.

Let us suppose that, as a result of asking and answering all these questions, our interpreter has satisfied himself that the indicated structure is genuine. Then he must ask what to do about migration.

In former days it used to be said that migration was not worthwhile if the dips were less than about 10 degrees. Our interpreter, however, is of the modern school, who hold that there are many advantages (in seismic-stratigraphic analysis, as well as in studies of structure) to migrating virtually everything.

At this stage our interpreter is heard to be muttering mutinously. The problem, apparently, is that his boss (for whom tidiness is a fetish) laid out the survey lines on a neat rectangular grid, without taking much account of the regional gravity maps, or the known structural grain, or the general fault orientation identifiable from satellite imagery. So the interpreter is now faced with lines which are neither on dip nor on strike, and which make small angles with most of the faults. Under these conditions two-dimensional migration is not appropriate. Further, the boss did not lay out the lines sufficiently close together to allow full three-dimensional migration. More mutinous mumblings. (What our interpreter does not know, of course, is that the boss had to lay out the lines along existing roads, because that was the only way of leaving himself enough budget to give a big increase of salary to his trusted and valued interpreters.)

In this case, we will say, no great harm is done; the reflections are easy to identify and to pick, the resulting contour map is therefore authoritative, and the migration can be done mechanically by migrating the contours. All is well, and the interpreter duly computes the closed area on the migrated contour map.

The closure, too. Of course, this is basically in milliseconds, but our interpreter is probably able to estimate (or even to guess) the local velocity just above reflector level to ± 10%, and so to convert into feet or metres with this accuracy.

With data of reasonable quality, then, and with bold and simple structure, the interpreter can feel that he has done a good and positive job, and that his estimates of closure and closed area are not likely to embarrass him too much when the structure is drilled. His greatest worry, probably, is the faulting. If the throws are small, important breaks in the reservoir continuity can exist without his knowing; the vertical resolution (that is, the seismic pulse shape) is inadequate to see the faults. If the throws are large, there is a risk that he has made the wrong correlations across the faults (so that the estimates of both area and closure are wrong). And if the seismic lines are too far apart, he may have been quite incorrect in assuming that *this* fault on one line was *that* fault on the next.

Yes, if he is good, our interpreter spends a long time agonizing about those faults. He goes to the geological library, and hopes that by reading between the long words he will be able to grasp some understanding of the tectonic history of the basin. He makes fence models of seismic sections, and squints into them from every angle in hopes of seeing the fault alignments. He checks and double-checks the contouring options introduced by different fault interpretations. And most of all he mutters mutinously against people who lay out seismic lines which are not at right angles to the fault grain.

When he has done his best with structure, he asks what the seismics say about the rocks.

From our discussion of seismic stratigraphy in Chapter 2, it is evident that the stratigraphic study of even a small part of the geologic column may require very extensive seismic coverage—preferably over the whole basin. This is where the big company really scores over the small. So our interpreter goes into the archives, to review the seismic data in other part of the basin.

He comes out sneezing...and boggled. The sheer immensity—even the mechanics—of the task defeats him. He concludes that such a project can hardly be contemplated until all the seismic sections and maps are in the computer's data base, summonable instantly (with consistent scales and formats) on a large-scale display.

Even without the practical problems, the problems of interpretation are not trivial. Our interpreter finds this as he commences to make what seismic-stratigraphic analysis he can, using just the present survey and last year's survey in an adjacent area.

As he looks at the data, his irritation with his boss gives way to something stronger: a positive scorn for authors who devise idealized illustrations like the seismic-stratigraphic one of Figure 21. For the extent of his prospect area is only a small part of the complete depositional system, from mountain-front to deep sea, and the analyses of Figure 21 are possible only by reason of the total view. Further, over his prospect the seismic-stratigraphic relationships are veiled by subsequent structure and distorted by subsequent faults. What he would like to do, first, is to flatten all structure, and unfault all faults, which occurred after the deposition of each geologic unit of interest—thus restoring each unit to the attitude and integrity which it had at the time of deposition.

Haltingly, tediously, our interpreter does it. (As it happens, there are no growth faults on this prospect; growth-fault areas go to the growth-fault specialist down the hall. But then, thinks our interpreter, if there were growth faults I would know what the depositional mechanism was anyway—and the lithology too. And although I could not unfault the faults, I could still draw the sequence boundaries containing each growth-fault system or succession.)

After the flattening and unfaulting at each level, the sequence boundaries and the sequence shapes stand out more clearly. Probably our interpreter does the exercise first on tracing paper, and is pleased to note that he now has much more confidence in the inferences from sequence shape. But if he wishes the same increased confidence in the inferences from reflection configuration *within* each sequence, he may find it desirable to have the processor produce a new section (one for each level) in which the flattening and unfaulting are actually done on the data—so that the reflection attitudes and continuity can be seen in the context of their depositional environment.

Perhaps our interpreter finds at one level a combination of sequence shape and onlap configuration within the sequence which suggests the presence of marginal-marine sands in the neighborhood of the present structural high. Then he has a specific prospect. This prospect is in part structural and in part stratigraphic. His problem now is to determine, from the onlap sequence and its details, the depositional strike of the sandstone bodies. To this he will later try to add some estimate of the sand thickness, or at least determine where that thickness is greatest. Then, having done what he can to *see* the sand body in its original condition, he will restore the structure and the faulting he had previously removed. From this emerges the first estimate of the best place to drill. Of course, this place is not necessarily the crest of the structure.

Or perhaps the evidence of the sequences suggests that the underlying structure was growing *during* deposition of a sequence of interest. Perhaps this growth served, fortuitously, to "stack" deltaic deposits which would otherwise have prograded basinwards—and so provided the prize of a stacked sequence of deltaic sands in a structurally high position.

Again, the interpreter would be seeking to establish the area and the thickness variations of the potential deltaic reservoirs—to visualize them just after deposition—and then to restore the subsequent structure and faulting to establish where the present prospects lie.

These exercises are also the basis of the interpreter's first estimates of the volume likely to be drained by a well. For example, we see in Figure 45 marginal-marine sands (identified as such by the onlap configuration of locally strong reflections on the unconformity) trending across and down the present superimposed structure. Apart from any later faulting, permeability barriers (in the form of shale breaks) are geologically likely between the individual sand lenses forming the gross sand body; our interpreter knows the attitude and orientation of these shale breaks from the general configuration of the sand. Probably he is not able to see the individual lenses or the shale breaks on his seismic sections; all he sees is the amplitude change at the onlap position, and as such the suggestion and general configuration of a gross sand body. But this is sufficient to tell him that he is dealing with individual reservoir sands which are elongated along depositional strike and of thickness less than wave base, and that the coarse stack

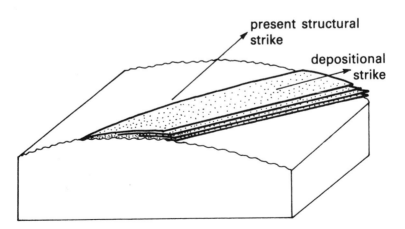

Figure 45 A series of shoreline sands subsequently deformed by structural activity with a different trend.

of sands which he sees is likely to be subdivided, in a reservoir sense, by shale breaks.

Clearest of all, of course, he knows that the area available for drainage, at the level of this sand, is not the closed area defined by the structure; at best it is the closed area of the shoreline sands.

From his seismic-stratigraphic analysis, then, our interpreter seeks to identify source rocks and reservoir rocks. For the source, he likes to see the low-energy reflection grain characteristic of marine shales. For the reservoirs, he highlights in particular shelf-edge conditions as the environment of reefs, shoreline conditions as the environment of beaches and bars, slope conditions as the environment of submarine fans, and oblique prograding-shelf conditions as the environment of deltas. Then he restores the subsequent structure and faulting to identify combinations of reservoir and trap—the habitats of oil and gas.

If our interpreter is asked to take his analysis further than this, he is likely to start grumbling about the data. You are going to have to pay for some fancy extra processing, he will say, and I must have at least two new lines. Shot in the right direction, he will add with heavy sarcasm; as we all know (huh), one line in the right direction can add more confidence than several in the wrong direction. And I would like to be consulted about the seismic source to be used, and the dimensions of the field layout, before these are decided.

So he gets his additional lines, well planned and executed, and he gets the remainder of the data reprocessed with more understanding of the geologic problems. In particular, he calls for the wavelet-processing steps we discussed in connection with Figures 34 and 35, for the sharpest pulse realistically obtainable (as discussed in connection with Figure 38), and for processing which preserves the validity of the indicated strength of the reflections. It costs some money, but by this stage his boss can see that he almost certainly has a drillable prospect.

The first thing our interpreter does, when he receives the new sections, is to search for fluid contacts. (Usually this means gas-liquid contacts, because the seismic contrast between oil and water is so poor.) A gas-liquid contact observed on a seismic section is of the *highest importance;* not only does it give a definite indication of gas (and so spit in the eye of those who love to say that gas is found only by the drill), but it also indicates that the porosity must be significant as well.

The gas-contact reflection arises because a rock with gas in the pores is softer (seismically speaking) than the same rock with liquid in the pores. The greater the porosity, the greater the contrast of hardness—and so the stronger the reflection. Further, the reflection is *always positive.* With the display convention which our interpreter uses, this means that a gas-contact reflection always appears black.

Gas contacts are usually horizontal in depth; in the presence of lateral variation of velocity, they need not always be horizontal in reflection time. So our interpreter scans his sections very carefully for reflection events which are *horizontal or near-horizontal, discordant against a background of dip,* and *black.* He calls these "flat spots." Figure 46 is an example.

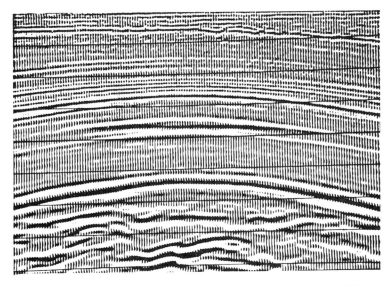

Figure 46 A "flat spot" representing a gas-liquid contact. (Courtesy Seiscom Delta Report)

Flat spots are not as certain as death or taxes, but they beat any other method of finding hydrocarbons.

Next, our interpreter looks for any local change of reflection strength which does not appear to owe its origin to the geology. He calls these "bright spots" and "dim spots."

For example, he may see a reflection, which he has previously identified as that between a massive low-energy shale and a massive low-energy carbonate, decrease in strength over a structural uplift. Is it a geometric effect of defocusing from a convex mirror? He must do the calculations, and decide. If not, is it because the carbonate has been fractured over the uplift (thereby becoming seismically softer), while the plastic shale has just adjusted? Perhaps. But if the diminution of reflection strength is very marked—if the dim spot is very dim—the message is that the hardness of the carbonate has been reduced almost to that of the shale, and this probably requires not only secondary porosity but also gas. So a dim spot on a structurally high and otherwise-strong carbonate reflection is an encouragement to expect gas.

For the same reason, our interpreter does not expect to see a very strong reflection from the top of a porous reef. Indeed, the chances of good porosity are better if no detectable reflection is observed, so that the reef is seen only as a blank in the continuity of the flanking reflections (as in Figure 21).

Bright spots, on the other hand, are usually associated with *sand* reservoirs—very porous, and containing at least some gas. In a sand-shale sequence, the shale-to-sand reflection is ordinarily positive and of medium strength for liquid-saturated sands of poor porosity. It becomes weak or even zero if the sand has good porosity;* this is just saying that a very porous sand may be no harder than a shale. But if the sand is very porous and contains gas, it may be significantly softer than the shale, and this may generate a very strong negative reflection—a bright spot. So the presence of a very strong reflection in a sand-shale sequence, where the geology would

---

*We note the conclusion: that there must be countless thousands of porous oil sands which do not show on seismic section.

have led us to expect a weak or medium reflection, may be a valuable and quasi-direct indication of gas.

Further, we have said that, with our interpreter's display convention, the central maximum of a positive reflection appears black, while for a negative reflection it appears white. A thin water-saturated sand (of the type which we considered in Figure 38f—reproduced as Figure 47) therefore appears as a black followed by a white; a thin gas-saturated sand (same figure) appears as a strong white followed by a strong black. Our interpreter likes to have both the strength and the white-black appearance, to be confident he has a classical bright spot (Figure 48).

Classical bright spots are most common in fairly young sand-shale sequences, at shallow to medium depth. They abound in the Gulf of Mexico, offshore Nigeria, and certain parts of southeast Asia. They always mean that some gas is present, but unfortunately a sand with just a few percent of gas

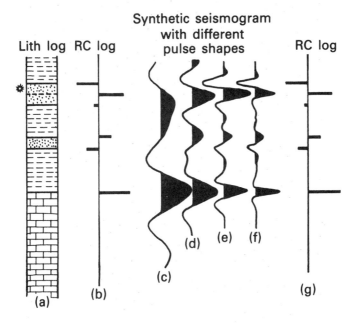

Figure 47 Reprise of Figure 38.

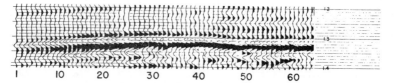

Figure 48 The "bright spot" response of a gas-saturated sand.

saturation yields a bright spot much the same as one from commercial gas saturation. In favourable cases the distinction can be made (as we shall see in a moment), but in general we have to admit some risk in drilling bright spots—not enough to discourage us, but enough to keep our fingers permanently crossed.

Let us suppose that our interpreter is working in an area young enough and with sand porous enough for bright spots to appear, and that he finds one. He finds the white-black gas-sand response of Figure 47f, and he finds it with very strong amplitudes. He identifies the white as the top of the gas sand, and the black as the base of the gas (which may be the gas-liquid contact, or the base of the sand itself). Since sand bodies (and gas-saturated zones in sand) tend to be lenticular in section, he may well see evidence of thickening and thinning within the bright spot. Perhaps one side of the bright spot looks like Figure 39 (here reproduced as Figure 49). In that case he jumps up and down with delight, for he can see the transition from a thick sand (with substantially separated reflections from top and base), through the tuned condition, to thinness and weakness. From the white-to-black time in the tuned condition he makes his first estimate of the thickness of the gas at that place (in milliseconds, of course). To the right of this he estimates the thickening by following the time between the same white and the same black, as they separate. To the left of it, he estimates the thinning by studying the decreasing amplitudes. Knowing that the sand must be porous and contain some gas, to give such a clear bright spot, he also knows that the range of values for velocity in the sand is small and low; he takes a value of 2100 m/s (7000 ft/s) knowing he cannot be far wrong. Then he converts the gas thickness from milliseconds to feet or metres, produces a contour map of gas thickness, and calculates the total volume of the gas-saturated sand.

That evening he surprises his wife by coming home in a good temper, and taking her to dinner.

To get more than this from the seismic data, the thickness of the gas must be sufficient to give clearly separate reflections from top and base. Unfortunately, at the present state of the art, this usually needs a sand of at least 150 ft (say 50 m)—and that eliminates individual sand bodies whose thickness is defined by the wave base. However, where sands have built up in a stillstand, or been deposited in dunes, the possibility exists of obtaining thickness, porosity, and a first estimate of total reserves, using the techniques of Chapter 3.

Two other things recur frequently to our interpreter's mind. One is that he must constantly be on the watch for indications of fracture (in the broken appearance of parts of the section, corroborated by local depressions of velocity), because he knows the importance of local fractures in increasing the appeal of a reservoir. The other is that his old so-and-so of a boss insists that his finished interpretation (whether structural or stratigraphic, or both) should be modelled back into a seismic section and compared with the original data. Our interpreter knows in his heart that this is a good pratice (and he will insist that *his* gorillas do it when he is the boss), but it certainly is a pain.

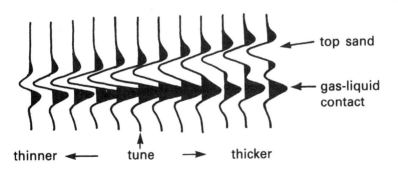

Figure 49 Reprise of Figure 39.

5
E.G.

The following pages give illustrations (Figures 50-81) of real seismic data. Some examples are old and poor; others are modern and excellent. (One or two are old and excellent!) The captions illustrate some of the thoughts which would be in the mind of the interpreter as he studies the data.

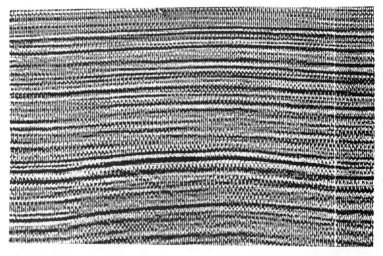

Figure 50 A simple dome structure. Look for evidence of this succession of events: deposition, uplift locally above sea level, erosion, rise of relative sea level, continued deposition, and further uplift. (Courtesy United States Geological Survey)

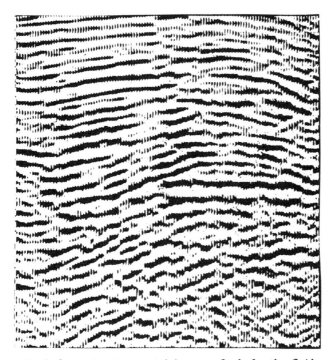

Figure 51 A dome structure containing gas. Look for the fluid-contact reflection. (From Bone; courtesy Geophysical Service Inc.)

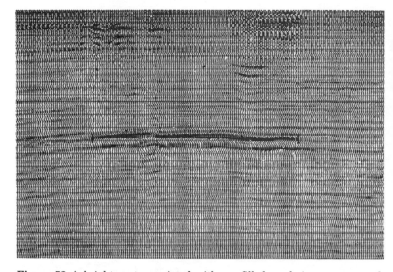

Figure 52 A bright spot associated with gas-filled sands in a young sand-shale sequence. Look for evidence of permeability barriers, and for places where the thickness of the reservoir could be studied by identifying the "tuning" of Figure 49. (From Larner; courtesy Western Geophysical Company)

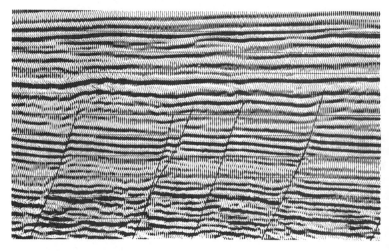

Figure 53 An example of normal faulting. What can we say about the age of the faulting? Does it depend on the lithologies present? (Courtesy Merlin Geophysical Company, Ltd.)

Figure 54 A complex bright spot. The interpreter must try to decide whether these are multiple gas sands en echelon, probably separated by near-horizontal shale breaks, and/or whether the reservoirs is cut by faults. (Courtesy Geoquest International, Inc.)

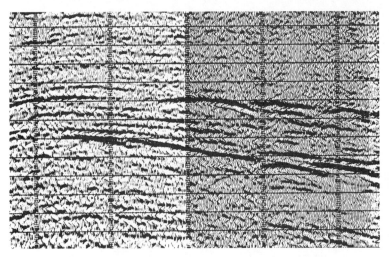

Figure 55 A probable stratigraphic trap; look for marginal-marine sand pulses, en echelon, and estimate the sequence of events, the relative sea level, and the abundance of sediment supply during deposition. What is the risk of error introduced by faulting? (Courtesy United States Geological Survey)

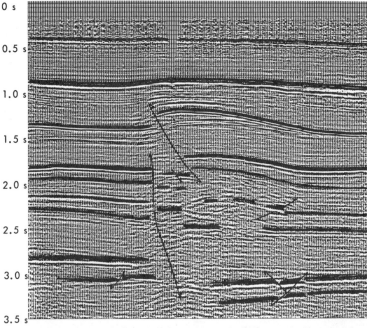

Figure 56 An interpreted section showing a variety of tectonic circumstances. Are there alternative interpretations at the deeper levels? What are the possible reasons for the local brightness at 1.1-1.2s, above the uplift? (Courtesy Prakla-Seismos Report)

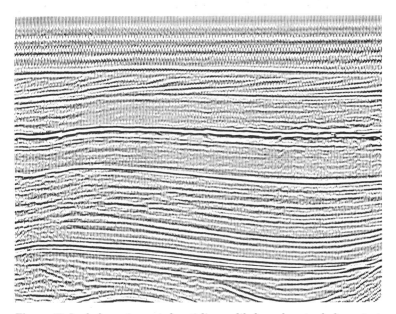

Figure 57 Look for: a truncated anticline, a likely carbonate-shale contact, shallow-water deposition, a prograding delta, minor faulting. What is the inference from the shift of the anticlinal axis with depth? (Courtesy Prakla-Seismos Report)

Figure 58 A strike line across a large submarine fan. Look for the gas-liquid contact. (Courtesy Seiscom Delta Report)

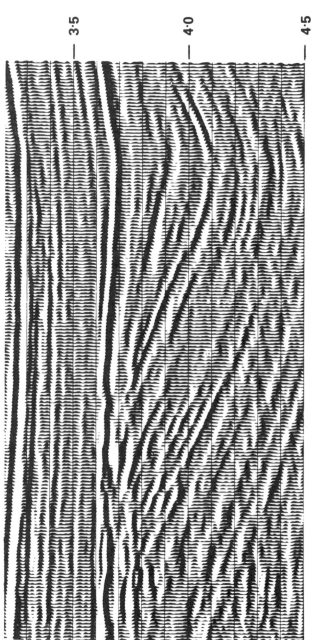

3·5

4·0

4·5

Figure 59 A major angular unconformity. (Courtesy Horizon Exploration, Ltd.)

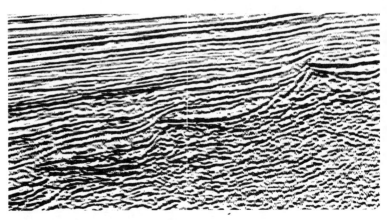

Figure 60 A fault terrace, showing fans derived from the right-most scarp overflowing down the terrace, and subsequent marine onlap on to the fan. (Courtesy Merlin Geophysical Company, Ltd.)

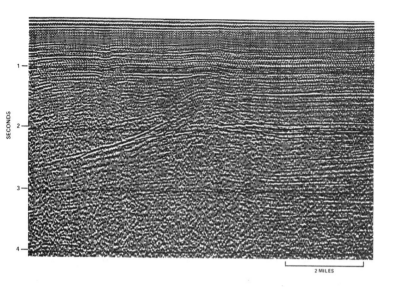

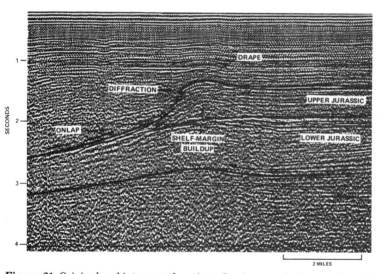

Figure 61 Original and interpreted sections showing a probable reef. Note in particular the position at the edge of the shelf, the drape above the feature, and the development of back-reef sediments. (From Bubb, Hatledid; courtesy AAPG)

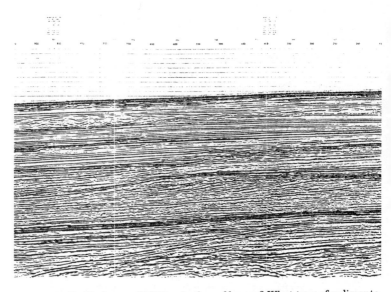

Figure 62 Shelf-edge reef? With a back-reef lagoon? What type of sedimentation below the lagoon? Elsewhere: a prograding channel? Deep-sea sands? (Courtesy Merlin Geophysical Company, Ltd.)

Figure 63 Look for: salt, a regional unconformity, calm-water deposition, an erosional channel. When did the salt start moving? When did the salt finish moving? (Courtesy Seiscom Delta Report)

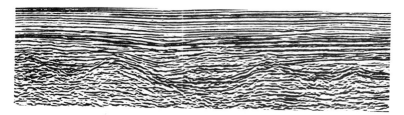

Figure 64 How many types of reservoir/trap situations are present here? (Courtesy Merlin Geophysical Company, Ltd.)

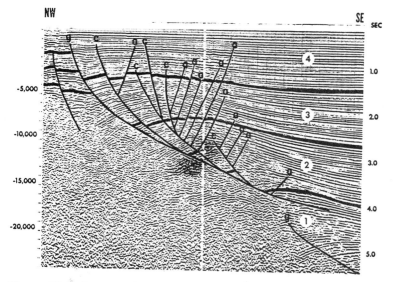

Figure 65 An interpreted section showing a major growth fault (g), the rollover on the downthrown side, and the associated crestal faults (c) and antithetic faults (a). (From Bruce; courtesy AAPG)

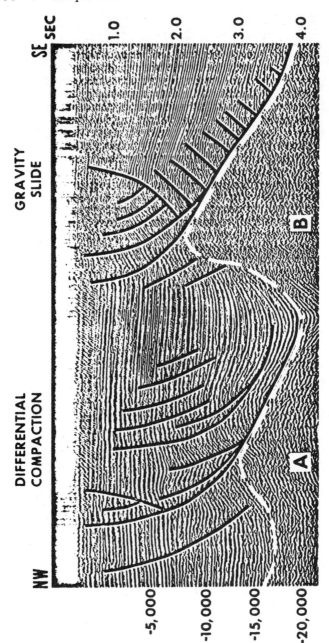

Figure 66 Another growth-fault example, showing the distinction between fault patterns introduced by differential compaction and by sliding under gravity. (From Bruce; courtesy AAPG)

Figure 67 An example to illustrate that growth-fault interpretation is much easier if someone else puts on the lines first. (Courtesy Seiscom Delta Report)

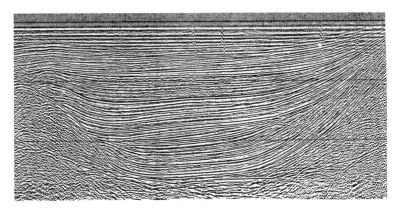

Figure 68 What is the inference from the eye-shaped feature just above two seconds, in terms of the time of salt movement? (Courtesy Prakla-Seismos Report)

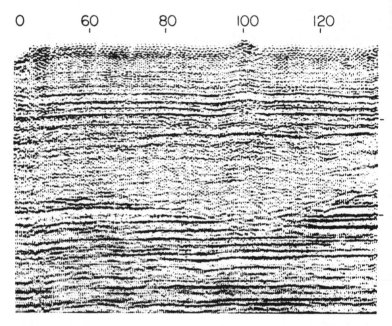

Figure 69 A clear channel, but filled with what? So where is the potential for a trap? (From Farr; courtesy Western Geophysical Company)

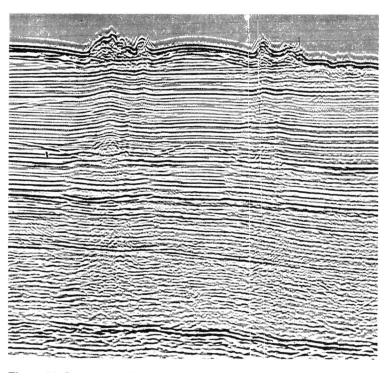

Figure 70 Gas vents producing mud volcanoes at the sea floor. What would be their effect as seismic lenses? What is the strange event cutting across the layering above the centre of the illustration? (Courtesy Merlin Geophysical Company, Ltd).

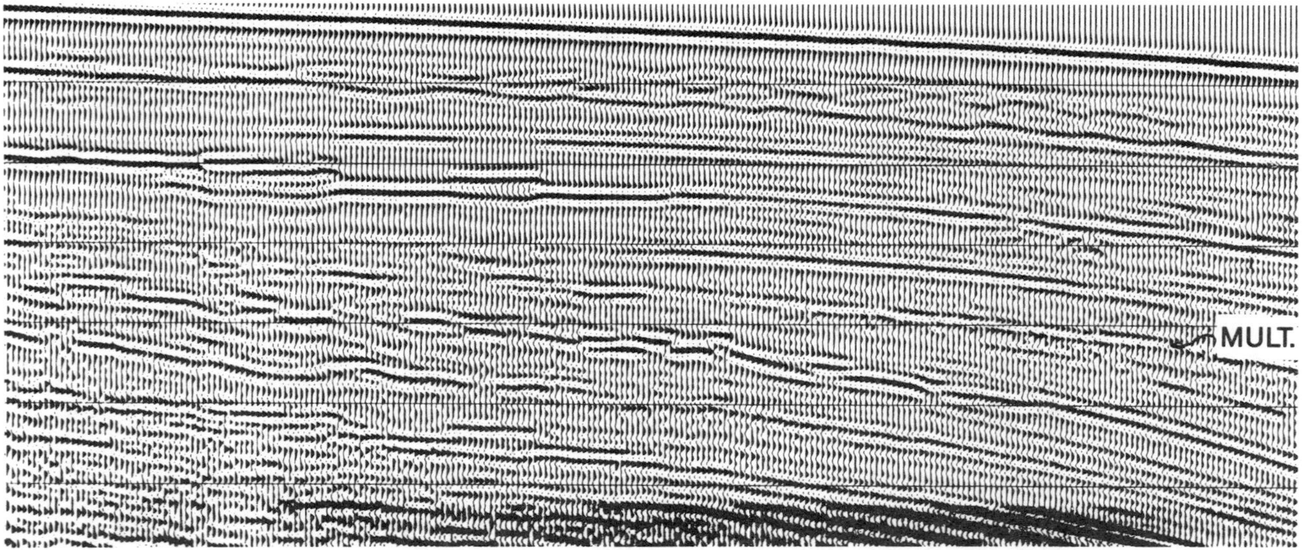

Figure 71 A section close to the present edge of the shelf. What have been the depositional regimes in different parts of the section? What would we expect to emerge if we could see in plan the features about 0.2s below the sea floor?

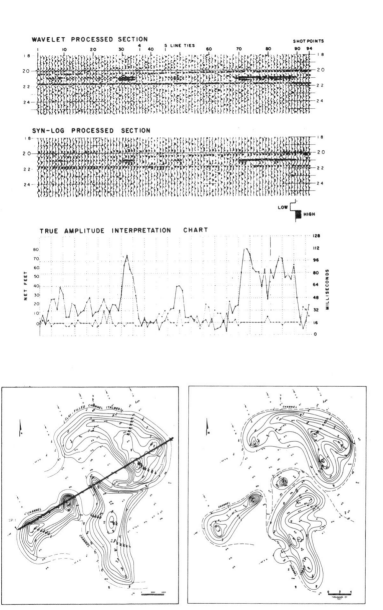

Figure 72 A discontinuous reflection occurs at 2.1s on the upper section. Its amplitude is plotted below. At bottom left we see the amplitude contoured from many lines (including the illustrated one as a heavy line), and interpreted as variations in the thickness of a group of point-bar sands. At bottom right we see the substantial confirmation from well control. (From Lindsey et al., May 1978; courtesy World Oil)

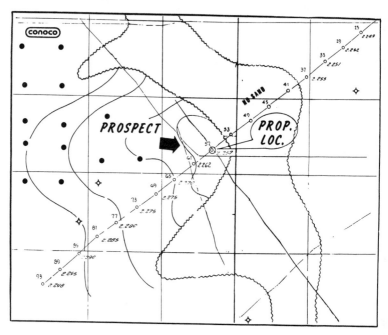

Figure 73 Figures 73-75 represent a prospect formed by a west-dipping reservoir sealed by a clay-filled channel. The channel, presumably a river and its tributaries, lies between the wavy lines. The reservoir attitude is shown by the contour lines. (Courtesy Conoco, Inc.)

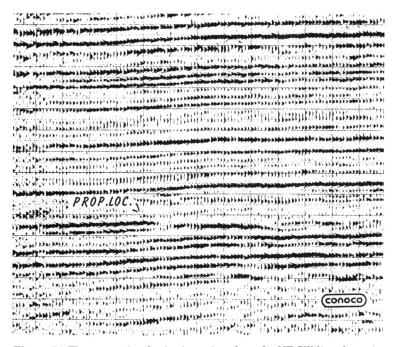

Figure 74 The conventional seismic section along the NE-SW line shown in Figure 73. The termination of the reservoir is clear. That the feature to the right of it is a channel follows more from its mapping in plan then from its direct appearance on the section. (Courtesy Conoco, Inc.)

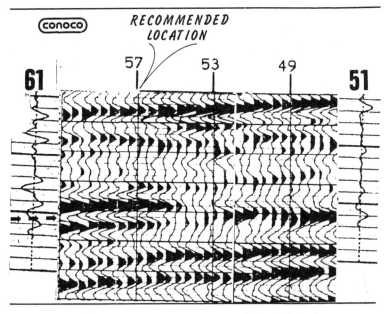

Figure 75 Detail of the zone where the channel seals the reservoir. At the left margin is a synthetic confirming the presence and interpretation of the reservoir; at the right margin is a synthetic doing the same for the clay-filled channel. (Courtesy Conoco, Inc.)

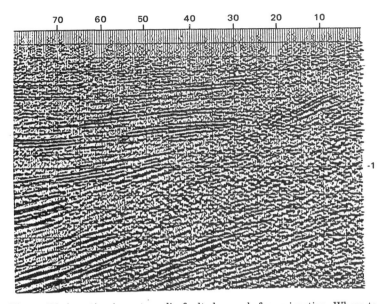

Figure 76 A section in a steep-dip faulted area, before migration. Where to put the faults? There is a potential fault trap at 1.6s under SP 55—would you drill there?

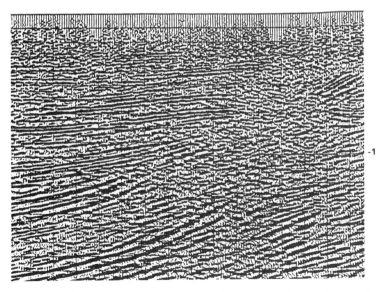

Figure 77 The section of Figure 76 after migration. The drilling location for the potential fault trap is now seen to be at SP 35, at a time of only 1.4s.

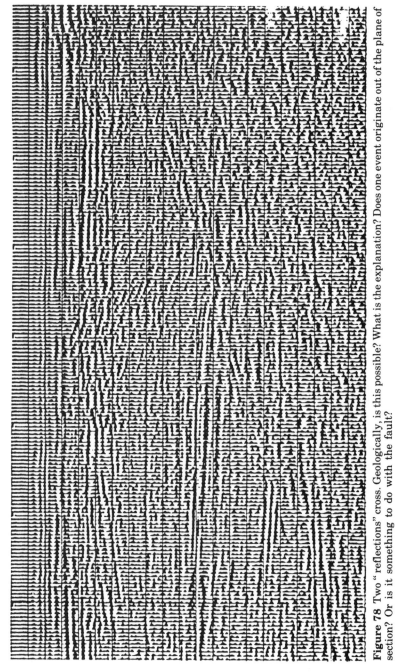

Figure 78 Two "reflections" cross. Geologically, is this possible? What is the explanation? Does one event originate out of the plane of section? Or is it something to do with the fault?

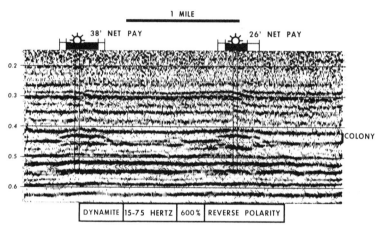

COLONY CHANNEL

Figure 79 An illustration where the channel is the reservoir rather than the seal. What depositional environment would we expect to infer from a line at right angles to this one? (From Focht, Baker; courtesy Hudson Bay Oil and Gas, Ltd.)

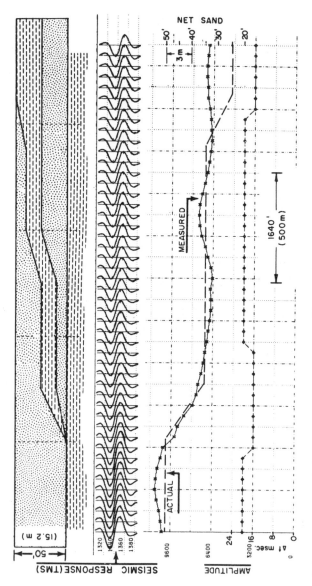

Figure 80 A remarkable illustration showing that the amplitude of the seismic reflection obtained from a thin sandstone containing a variable thickness of shale is closely proportional to the *net* thickness of sand. (From Schramm et al.; courtesy AAPG)

Figure 81 An example to illustrate that other people have problems too. How to correlate across the fault?

6

Borehole Seismics

We have stressed in previous chapters that raw sections obtained with surface seismics are *time* sections. If we do not know the velocities, our reflection identifications—the correlation of reflection events with geologic markers at known depth—involve an element of risk. We also agreed that variations of velocity from place to place jeopardize our structural interpretations.

Where we have wells in the area, the answer is obvious: lower a geophone down the hole, make a bang at the surface, and see how long the bang takes to reach known depth. A hurrah for simplicity.

This operation is called a check-shoot (or, formally, a velocity survey; Figure 82). It may be done in cased or uncased holes, using a logging cable or a separate wireline, with the rig in place or from a simple A-frame. The most important point is to use a motion-sensitive (rather than pressure-sensitive) geophone, and to be sure that it is locked to the wall of the hole. In cased holes it is sufficient to use a strong leaf-spring; the geophone is lowered to total depth, the spring is actuated, and the geophone is dragged upward from one recording position to the next. In uncased holes it is necessary to use one (or preferably two) remotely actuated locking arms or pistons, which can be extended to force the geophone against the wall of the hole. When the geophone is locked, the cable is slackened and the system allowed to stabilize. When all of this is properly done, the geophone is remarkably quiet—down there in the body of the earth—and we shall see later that this offers a significant advantage over a geophone on the surface.

In former days, the bang was usually a single charge of explosive in a shallow hole. As such, it had to be spaced some distance from the wellhead, to avoid risk of damage; the necessary horizontal offset introduced a minor uncertainty into the results. Today it is usual to employ several or many small bangs closer to the wellhead, and to add together the resultant signals until sufficient clarity of the downhole arrival is obtained. In this case the bangs may come from small explosive charges (for example, 100 g or ¼ lb), or from air-guns

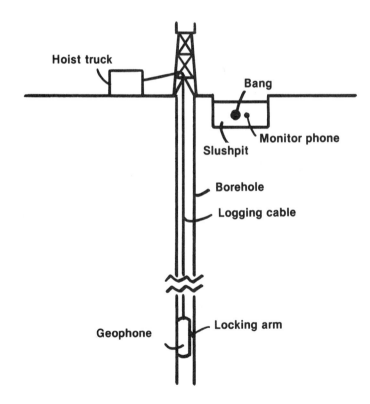

Figure 82 The basic arrangement for a velocity survey in a borehole.

in the slushpit or the sea, or from various forms of mechanical hammer. Or, of course, we may use Vibroseis; we shall talk about this later.

The operation then consists in positioning the geophone at several selected locations (normally just below the top of important marker formations) and actuating the bang—or stack of bangs—at each such location.

A typical result is shown in Figure 83. The vertical scale represents depth in the hole. The trace obtained for each geophone position is plotted at the appropriate depth, with travel-time increasing to the right. Then a line joining the arrivals on successive traces is a graph of travel-time against depth; from this we can plot the velocity as a function of depth (Figure 84). We can also compute the interval velocities

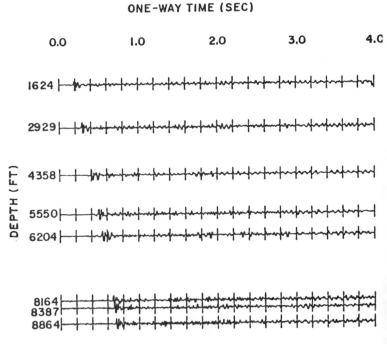

Figure 83 Results from a basic velocity survey. (From Lang; courtesy Geosource Inc.)

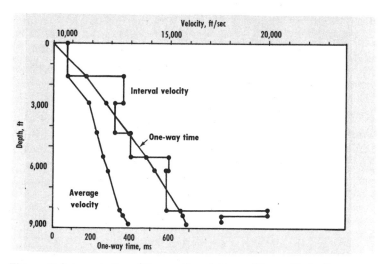

Figure 84 Interpretation of the results: relations of time with depth, average velocity with time, and interval velocity with depth. (From Lang; courtesy Geosource Inc.)

between geophone positions; however, these have no lithologic significance unless the geophone positions actually coincide with all the formation tops at which significant variations of velocity occur.

The value of all this for reservoir studies is indirect. True, it gives us more confidence in our structural interpretation and in our closures. True, it gives us the exact time at which the fluid contact exists on the seismic section (even if it is an oil-water contact and we cannot see it); this improves our estimate of the area of the hydrocarbon accumulation. True, it allows calibration of the sonic log (forcing the integrated sonic to agree with the check-shoot), and thus adds further assurance to the synthetic seismograms we discussed in Chapter 3. But to bring new information to studies of the reservoir, we need to develop the check-shoot into a *vertical seismic profile (VSP)*.

Vertical seismic profiling is conceptually and operationally the same as check-shooting, except that more geophone locations are used (typically at a regular spacing of 30 m or 100 ft—or even less), that the recording continues for some time after the direct bang has arrived, and that we pay rather more attention to the form of the bang. The display then appears as in Figure 85.

Obviously, we can see the direct arrival of the bang even more clearly than before. But now we can also see reflections of the bang—originating at the depth of major contacts, and slicing upwards across the display with a velocity the same as that of the direct downgoing arrival. What is more, we also see multiple reflections—originating at the underside of major contacts (particularly the surface) and slicing downwards again. Of course, geophysicists are tickled pink by this—it is such a *graphic* demonstration of the reflection process.

By this stage of the book all of us—geologists, engineers, everybody—are very comfortable looking at seismic sections...squinting along the reflection alignments and grunting. Now we must all learn to do the same for a VSP.

Figure 85, although an old example, is quite a good one. We can see the direct downgoing arrival even without squinting. The three marked upcoming reflections do not need too much squint, and we can probably see some other ones as well. The marked downgoing multiples are also clear. And at this stage we have no difficulty with the format of the display; the vertical axis is depth down the hole, and the horizontal axis is seismic travel time.

Figure 86, however, is a poor VSP in that a totally spurious wave-train cuts right across it at low velocity. This is the "tube" wave—the bang propagating in the mud of the hole rather than in the formations. It is caused by a poor geophone-locking mechanism, or by having the bang too close to the wellhead. For check-shooting, of course, it would be immaterial; it is only when we need to see the reflections, coming in after the direct arrival, that we must take great care to suppress the tube wave. If the service company sends us a VSP like this, we would send it back.

Figure 87 is a better one, though a tube wave can be seen at shallow depth. In fact, the whole of the shallow portion is noisy; usually this is unavoidable for very shallow positions of the geophone, but in the present case it probably suggests a locking problem, or a poor cement job, or the swaying of the

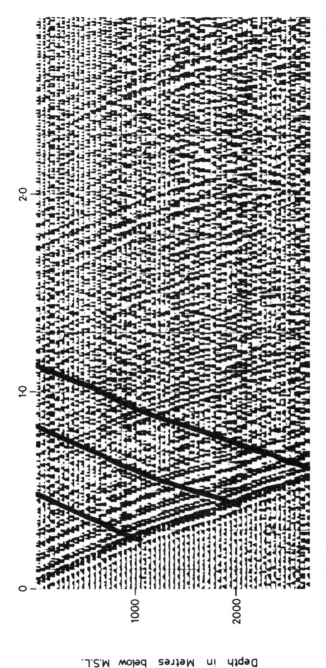

Figure 85 An early VSP. (Courtesy Seismograph Service Limited)

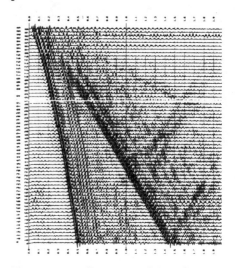

Figure 86 An illustration of the tube wave.

service rig in the wind. Incidentally, we notice at the left how the direct-arrival times and the spacing of the geophone positions can be manipulated to give us rather detailed interval velocities.

Figure 88 is a good modern VSP. Our immediate reaction is to notice the dominance of the downgoing multiple reflections over the upcoming reflections. The message must be that the shallow part of the section (above the shallowest geophone position) must contain some very strong reflectors; the bang rattles up and down in these shallow layers, and builds up a long reverberant tail to the downgoing signal.

But we can still see several upcoming reflections (notably at BC and DEF). Again, as always, we can see a reflection more clearly if we squint along the alignment. What are we doing when we take this foreshortened view? We are optically mixing together, or stacking, many traces along the alignment we choose. Now this is something the computer could do for us, very simply. For horizontal reflectors we know that the alignment of the upcoming reflection is the mirror-image of that of the downgoing direct arrival; therefore all we have to do is to feed this information into the machine, and tell it to work

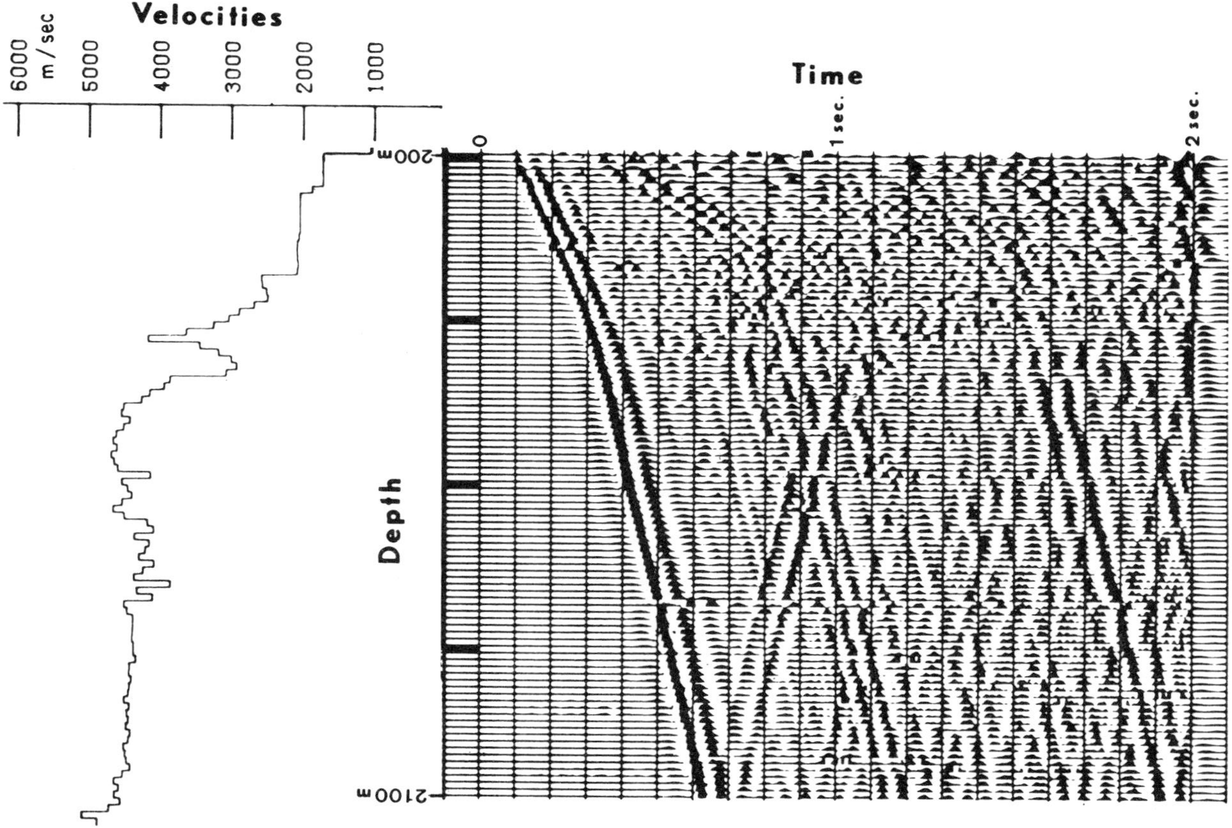

Figure 87 A VSP with its interval-velocity interpretation. (Courtesy Compagnie Générale de Géophysique)

111

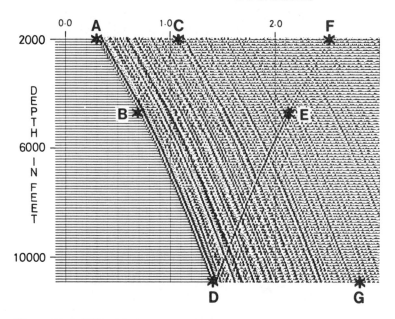

Figure 88 A VSP with very strong downgoing reverberations. (From Hubbard; courtesy Seismograph Service Limited)

its way along the alignment summing some suitable number of traces. This emphasizes the upcoming reflections, just as we do by squinting along them.

In Figure 89 we see another VSP, on which the upcoming reflections have been emphasized in this manner, and on which the untreated direct downgoing signal has been merged back into the display. This is a beautiful VSP indeed.

The topmost trace in Figure 89 is very shallow. Let us now squint along the body of upcoming reflections, and project them up to a line representing the surface. What do we have?

We have the surface-to-surface seismic trace—the one we should get on a normal seismic section as it passes through the well. In a sense, then, the VSP is allowing us to construct a projected trace which has the same property as the synthetic trace—that of predicting what the surface reflection method will obtain.

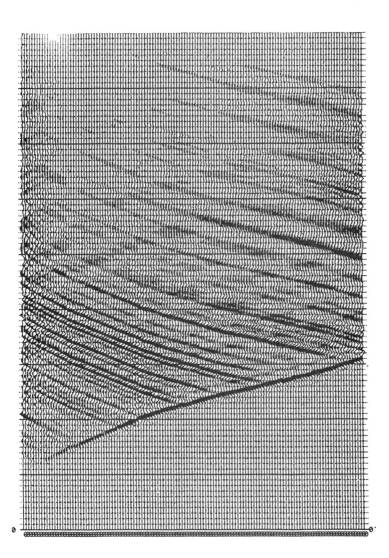

Figure 89 A VSP with processing to enhance the downgoing direct signal and the upcoming reflected signals. (From Balch; courtesy United States Geological Survey)

The synthetic trace is subject to imperfections in the sonic and density logs, and to errors if the pulse shape selected is not correct. Consequently, we are not surprised that the VSP-projected trace usually does better than the synthetic in giving a match to the actual trace obtained with the surface method.

The technique of selectively enhancing the upcoming waves leads to a different form of display for the VSP. Figure 90 is this display for the data of Figure 88. In addition to applying the selective enhancement, the computer has delayed each trace by the time of the direct downgoing signal. It looks rather like a normal surface-to-surface section. However, it is not; the horizontal scale is depth of the geophone down the hole.

If a reflection is not horizontal on this display, there is dip at that level. There have been examples in which the operator was alerted by the VSP to the presence of a nearby salt dome, or a nearby fault plane. However, the VSP remains basically a one-dimensional observation; it may tell us there is dip, but it does not tell us the direction. For that, we must take the next step.

The next step depends on the circumstances and the problem. Suppose that a wildcat drilled solely on subsurface data is dry, but that it encounters near-reef material at medium depth. A reef is near, but how near? And in which direction? The best approach to this problem may be several normal surface-to-surface profiles radially through the well, coupled with a synthetic seismogram (calibrated by check-shooting) to establish the time at which the reef indications can be expected. Figure 91 might represent such an example; it illustrates one line (of three) through the well, split to show the synthetic and electric logs at the well location.

Sometimes, we see, nature is kind; she gives us two reefs where we dared to hope for one. More often, of course...but then the example would never have been shown.

Or suppose that an offshore field is being developed by directional drilling from a central platform, and that the reservoir exhibits unexpected changes between the central well and one of the deviated wells. The problem is to establish where the change occurs, so that further wells can be deviated

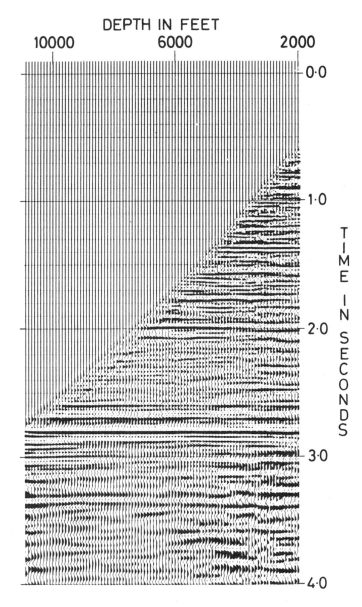

DEPTH IN FEET

10000   6000   2000

TIME IN SECONDS

0·0

1·0

2·0

3·0

4·0

UPGOING WAVES

Figure 90 Selective enhancement of the upcoming reflections of Figure 88. (From Hubbard; courtesy of Seismograph Service Limited)

115

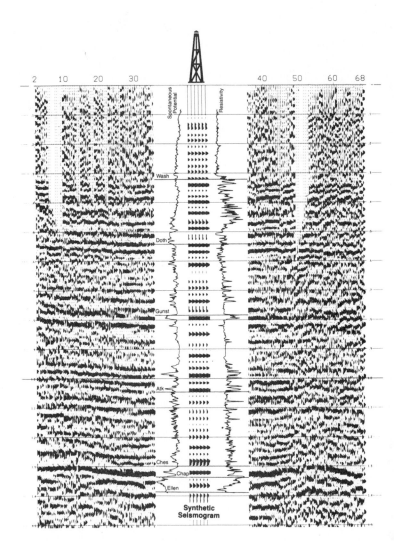

Figure 91 A surface seismics profile through a well which encountered near-reef material. Two reefs are evident, with the well between them. (From Lang; courtesy Geosource Inc.)

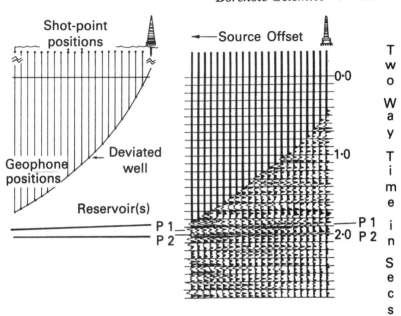

Figure 92 A VSP in a deviated hole, processed to enhance the upcoming reflections from the reservoir. (From Kennett; courtesy Seismograph Service Limited)

appropriately. Figure 92 illustrates a solution: a VSP in the deviated hole, with the air-gun source in a boat maintained vertically above the geophone. As it happens, this particular example shows the reservoir to be very well behaved (within the limits of seismic resolution).

Or suppose a more general situation, with a single vertical well, in which the problem is to determine the reservoir limits imposed by faulting, or by pinchout, or by loss of porosity. Then the solution is a *3-D VSP*.

The 3-D VSP is obtained with a mobile source, shooting from many surface positions into a borehole geophone (Figure 93). Again the source positions may be disposed along three lines at 60-degree intervals through the wellhead (Figure 94); detailed problems may require more than three.

The source may conveniently be an air-gun array, or a weight-drop, or one or more Vibroseis vibrators. In any case,

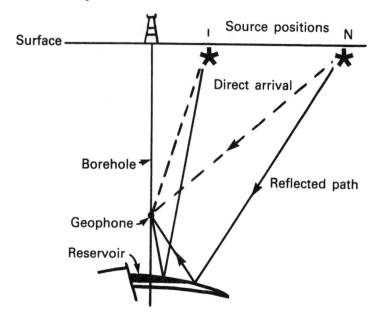

Figure 93 The principle of the 3-D (or offset) VSP.

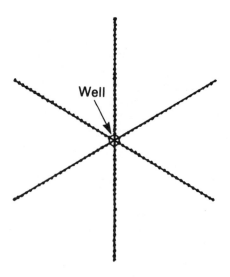

Figure 94 Plan view of radial lines of source positions as used for a 3-D VSP.

the recording equipment allows the summation of several emissions from each source position, so that adequate signal can be ensured.

For one position of the geophone in the hole, and for each line of source positions, we obtain traces in cross-sectional format. In idealized form, and for just a few traces, they might look like Figure 95. We see the direct arrival from source to geophone (with its characteristic normal-moveout shape), followed by a reflection complex derived from the top and base of the reservoir. In particular, we see at the right-hand side evidence of reflection amplitude and shape which suggests the thinning out of the reservoir, while at the left-hand side we see the limit of the reservoir imposed by the fault. By relating the equivalent information on other lines through the well, we hope to delineate the limits of the reservoir body rather closely.

If we have gas, we would also hope to see the fluid contact, to define its limits, and to identify breaks in its continuity. If

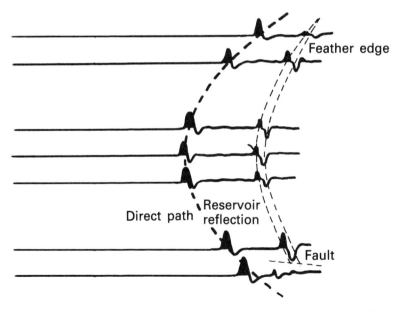

Figure 95 An idealized result from a 3-D VSP line, corresponding to the sketch of Figure 93.

we have oil, our chances of this are much smaller, but there is still a possibility that zones of locally improved porosity may still show as changes of reflection amplitude; to suggest this the central three traces of the figure show a smaller amplitude (meaning better porosity, for an oil-sand reservoir overlain by shale).

We note also that a 3-D VSP gives us much more confidence in our depths, and in the significance and accuracy of our reflection amplitudes, than we obtain with surface seismics; that is because we always have a record of the direct bang as it passes the geophone on the way down. We can calculate reflection coefficients directly, by isolating and relating the reflected upcoming bang to the direct downgoing bang. Further, intriguing new possibilities arise in using knowledge of the form of the downgoing bang to process the reflections.

Figure 96 gives an illustration of a reservoir study by a 3-D VSP. Of course, no one would attempt a full interpretation without the other lines, and it is clear that problems exist with the reverberant tail of the downgoing pulse; nevertheless, first appearances suggest a fluid contact to the left of the well, a significant change in the P1 reservoir very close to the well, and a probable limit between the sixth and seventh traces to the right of the well.

At this stage it is evident that a strong case can be made for making VSP studies on discovery wells. However, we should acknowledge immediately that there is no consensus on this. In Russia, a VSP is run on "virtually every well; even development wells" (Galperin); the VSP may even be repeated, after a few years production, to check whether there are sealed-off parts of the reservoir not contributing to production. In the North Sea, VSPs are becoming standard on wildcats. In the U.S., two or three companies have developed special VSP equipment, which they move around from well to well under cloak of extreme secrecy. But in general it remains difficult for geophysicists to get approval to conduct borehole work, and contractors who were considering entering the business have decided to stay out. Why?

Let us review what benefits can accrue from borehole seismics, and then, at the end, let us ask the question again.

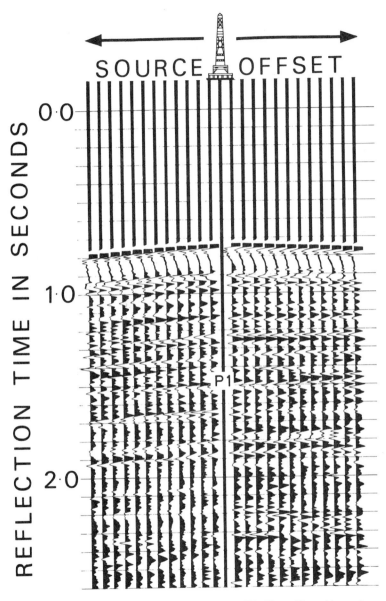

Figure 96 Delineation of a reservoir by a 3-D VSP. (From Kennett; courtesy Seismograph Service Limited)

(1)   The log analyst normally decides which porosity log he will specify on the basis of the lithologic and other conditions in the zone of interest. Of course he will continue to do this. However, we should record that the geophysicist's first choice, for his own specialized purposes, is always for the sonic log. The geophysicist also likes to have the density log, as a comforting (though usually minor) input to his calculations. The product of sonic velocity and formation density defines the acoustic impedance (or hardness), and so defines the ability of the geologic column to generate seismic reflections. From this product the geophysicist calculates the synthetic seismogram, which is an important tool in the interpretation of seismic data.

(2)   The vertical scale of the logs is depth, whereas the vertical scale of seismic data is time. Over short intervals, and where the reflection identifications are known without doubt, the calculation of the synthetic seismogram on a scale of time can be done using the integrated travel-time pips on the sonic log (preferably after skilled editing).

(3)   When a synthetic seismogram has been calculated from the logs, we know (within limits imposed by the quality of the logs and the uncertainty of the seismic pulse) what result we expect to obtain from surface seismics across the well location. First we check that this is indeed obtained—that we have a match with the field data. Then we perturb the logs at the reservoir level—thickening or thinning or otherwise changing the reservoir—and calculate the corresponding synthetics. Finally we seek a match between the perturbed synthetics and the seismic data away from the well, in hopes of drawing conclusions about the lateral extent and variability of the reservoir. Provided that we already have a seismic line or lines across the well, this is easy and inexpensive.

(4)   However, if the synthetic is required over a long interval, and/or the reflection identifications cannot be made without doubt, the conversion of the logs

from depth to time cannot be made from the integrated sonic pips. It is necessary to calibrate the sonic by check-shooting. Two benefits in addition to the calibration are obtained. The first is that check-shots in a number of wells in an area give authoritative velocities for conversion of seismic sections to depth, establish the lateral variation of velocities over the area, and so increase confidence in seismic indications of structure between the wells. The second is that the interval velocities over long intervals in the well can be related to the lithology in the well and to the interval velocities measured from surface seismics at the well, so that predictions of lithologic change can be made by studying the changes in the interval velocities measured away from the well.

(5)   At minor additional expense, the check-shooting (if done by a well-equipped crew) can be developed into a full VSP. The additional benefits which accrue are these:

- The VSP yields greater certainty in the velocities (particularly in the interval velocities).
- The VSP clarifies the origin of reflections (and multiple reflections). In particular, it establishes whether certain reflections visible on the normal seismic sections are time-stratigraphic horizons, or lithostratigraphic horizons, or contrasts associated with local cementation, local porosity, or the presence of hydrocarbons.
- The VSP allows selective enhancement of the upcoming reflections, and the projection of their alignment to the surface. This provides an excellent prediction of the best that surface seismics can hope to provide at that location. In conjunction with the synthetic, the projected VSP establishes whether a particular reservoir problem is soluble by surface seismics or not; it may therefore prevent the wastage of money on impossible problems.

- The projected VSP usually gives a better match to surface-seismic data than does the synthetic. However, it is not possible to perturb the projected VSP artificially, as one can do with the synthetic; consequently the use of surface seismics to indicate reservoir changes away from the well is less conclusive.
- The VSP can look ahead of the drill. It shows (in milliseconds) how much further the drill must go to encounter a particular reflector; it cannot, in itself, give this information in metres or feet.
- The VSP gives a measure of the dip of reflectors penetrated by or beneath the borehole (though it does not give the direction of the dip).
- It alerts us to the presence of a reflecting fault plane or a reflecting intrusion close to the borehole (though again it does not tell us in which direction).
- In areas of structural complexity, it yields reflections from reflectors which (because of the bending of seismic reflection paths) are not recordable at the surface.
- When the VSP provides a direct measurement of the interval velocity with a *liquid* saturated reservoir of known lithology, it provides its own measure of porosity. This is conceptually no different from the measure provided by the sonic log; however, the volume of rock represented by the VSP measurement is much larger than that sampled by the sonic. The VSP measurement may well be the more useful in predicting the performance of the reservoir.
- The VSP allows the direct computation of reflection coefficients. In conjunction with its indications of interval velocity, this allows the above computation of porosity to be realized for *gas*-saturated reservoirs also.
- The VSP provides a recording of the downgoing bang, which can be used to advantage in the

processing of the upcoming reflections. Specifically, the techniques of wavelet processing discussed in Chapter 3 become very strong indeed
with this advantage.

(6) If the seismic source used for the VSP is mobile, and
if appropriate access and permits can be obtained,
one or more positions of the borehole geophone may
be recorded with one or more horizontal profiles of
source positions (Figure 94). In the limit, this allows
a full three-dimensional study of the reservoir and
its limits. Within the bounds of seismic resolution, it
offers the ability to assess the thickening and thinning of the reservoir body; to establish its extent, dip
and orientation; to find local zones of increased or
decreased porosity; to establish the position of faults;
and (in the sense that it may attribute a lenticular
nature to individual reservoir bodies) to predict the
likely location of shale breaks within the gross
reservoir. It is thus highly material to the placement
of delineation and development wells. Further, it can
provide this information to depths well below the
present depth of the borehole. Further again, it
establishes the correct correlation between wells
whose conventional logs defy correlation.

All of these inducements are shouted very loudly by the
contractors offering 3-D VSP. "So what?" say the others—"We
can do all that by standard surface seismics, with a few lines
radially through the well." Certainly there is a measure of
choice, and—since the choice has to be made by exploration
people like ourselves—it helps to set down here the advantages and disadvantages affecting that choice.

Let us put VSP in the pillory first.

The major technical disadvantage of the 3-D VSP, relative
to surface seismics, is that it gives useful reflection information only below the geophone. If we have a number of possible
pay-zones, over a large range of depth, the geophone must be
well above the shallowest one. But this reduces some of the

advantages (listed later) of a deep geophone, making the results more like surface-seismic results.

Further, considerations of reflection angle limit the maximum extent of the reservoir which can be studied (Figure 97); if we assign some maximum angle, this defines the geophone position necessary for a reservoir of given extent. We shall see in the next chapter that there is some value in going to wide reflection angles for specialized types of study, but for simple and straightforward interpretation we probably prefer to keep the angle of the figure below about 60°. To a first approximation, this means that the horizontal distance from the borehole to the edge of the reservoir is about 0.6 of the vertical distance from the reservoir to the geophone. If the reservoir edge is 1200 m from the borehole, the geophone must be 2000 m above the reservoir—which requires that the reservoir be more than 2000 m deep.

We might say, as a simple rule of thumb, that 3-D VSP is most appropriate to the case where we need to study the details of the reservoir out to a distance of one-third of the reservoir

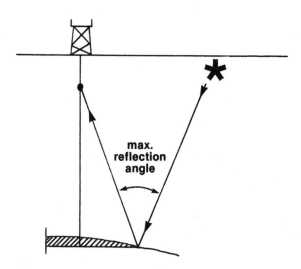

Figure 97 The maximum extent of the reservoir which can be directly tested by a 3-D VSP is limited by the reflection angle.

depth. Beyond this, it would be wiser to choose surface seismics.

Next, we note that the 3-D VSP requires dedicated use of the hole. If the well is in production the geophone may be run through the lubricator, but the well must be shut in. Surface work does not need the hole at all (though in practice it is sometimes necessary to shut down all the engines on the rig—for a period which may be longer than for a 3-D VSP).

Finally, the 3-D VSP may be more of an organizational headache than the surface seismics. This is because the surface-seismic crew comes as a self-contained entity, whereas a 3-D VSP may involve forcing a cooperation between a travel-weary VSP operator who is longing for some sleep and a reluctant driller who is apprehensive about his hole.

Now let us hear the advocates of the VSP.

There are operational and cost items on this side too. The equipment is cheaper, and its operation needs fewer men. Access and permitting problems are somewhat reduced. At sea it requires only a local boat outfitted with a transportable source system; it does not require a full (and expensive) seismic vessel. And the processing is less expensive. But it is probably in the potential for technical advantage that the final factor lies.

The VSP offers *reliable* velocity information, immune from the problems that afflict velocity estimation from surface seismics. As such it guarantees proper corrections, proper time-depth conversion and proper migration—all very important to reservoir studies. Near-surface problems—which contribute another risk of false structure—are halved. Further, the VSP data are likely to be *better resolved,* and hence better for the study of subtle features of the reservoir. This is so for several reasons.

- Surface-seismic data are usually compromised to some extent by the need to reduce unwanted types of waves (in particular, waves spreading out from the bang along the surface). These compromises, which express themselves as "arrays" of one form or another, tend to smear the indications of reservoir features. There is much less requirement for array smearing in a well-designed VSP.

- For VSP, the seismic bang need travel only once through the troublesome and absorptive near-surface zone. This improves our chances of keeping the reflected bang as a short sharp click rather than a dull rumbly thud, and so of seeing thin reservoir beds; it improves the vertical resolution.

- Improving the vertical resolution also improves the horizontal resolution—the ability to see edges, barriers and faults. This is improved still further in a VSP, because the total length of the seismic path (from source to reflector to geophone) is reduced; a shorter path length improves the vertical and horizontal resolution, both independently and conjointly.

After all this, the bemused professional is likely to ask whether he really must make a choice between 3-D VSP and surface seismics—is there no way he can avoid the risk of making the wrong decision.

Of course. Shoot both. All it takes is money. But the fact is that most of us (far from shooting both, which would be technically superb) do not even shoot either, and this brings us back to our earlier question—Why?

In part, it is because many of us are working in mature areas, where the reservoirs of current interest are too thin to be seen seismically—even with the benefits of the VSP. Only the new techniques of the next chapter offer us hope of seeing a 5-metre sand. That is why the VSP techniques are most active, presently, in the frontier areas. And even where the conditions are appropriate, there is still the problem of getting a VSP crew to the site; it is not yet like making a call to the local office of the logging company.

Further, we must acknowledge that those geophysicists who have some experience of interpreting VSPs know that it is not always easy, while those without experience have a natural fear of the unknown.

Further, many geophysicists work in exploration departments in which their responsibility ends sharply when the wildcat location is decided. They have no further involvement with the development of the field. They have been taught that rigs can never be shut down just for them, nor production shut in, and they accept it. They do not know enough of the technical and financial factors affecting field development to allow them to assess the relative cost-effectiveness of a 3-D seismic survey versus a delineation well, and so they keep quiet. And while they keep quiet, the other professionals in the team never learn that there is any alternative to the delineation well.

Which is one of the reasons for this book.

7

Current Efforts to Improve Seismic Delineation

Broadly, these current developments are aimed at:

- increasing the density of seismic control over a reservoir, to improve the definition of faults and other barriers in plan,

- improving seismic resolution, to see thinner beds and smaller features, and

- learning more about the nature of the rocks, by making additional seismic measurements.

We discuss these in turn.

7.1 3-D Surface Seismics

In Chapter 1, we discussed the old-established technique of shooting a *grid* of seismic lines, and of manipulating the results into a contour map. The spacing of the grid, obviously, must be smaller than the dimensions of any feature which we cannot afford to miss. In times past, this might have meant a kilometre or more; many areas of the U.S., for example, have been shot primarily on the section lines, at 1-mile spacing. But when we talk of reservoir delineation, the dimensions of any feature we cannot afford to miss become much smaller. In a detail situation, we must shrink the spacing of the seismic grid to 100 m, or perhaps even less.

In principle, there is nothing new in 3-D seismics—just the adoption of smaller grid spacings. At sea, the field work is almost indistinguishable; individual lines are shot as before, but the lines are much closer. On land, the field work can be adjusted to the surface conditions; in some areas it may be most convenient to build up the coverage in a series of swaths, while in others the best approach may be lines of source positions perpendicular to lines of geophone positions. Whatever the technique, the object is to build up a dense areal coverage of reflection points over the reservoir.

With that done, new possibilities exist in the processing— possibilities associated with the fact that we now have a volume of data points. The processing and utilization of the data can now be truly three-dimensional, without regard to lines of profile.

Most important, in this context, is the ability to perform a full three-dimensional migration. As we discussed in Chapter 1, this moves the reflection indications to their correct point of origin in three-dimensional space. Pinch-outs, faults and structural flanks now emerge in their correct positions and relationships, whatever the orientation of the seismic lines relative to the structural grain. This, as we agreed before, is the complete answer to our traditional apprehensions that things may not be exactly where we say they are.

Intriguing new possibilities also exist in the display of the migrated data. First, we can call for a standard cross-sectional display at any bearing, irrespective of the bearing of the grid lines. Second, we can call for a zig-zag cross-section, of the type which we need for tying a succession of wells not in a straight line. This is an altogether new luxury. Third, we can arrange closely-spaced cross-sections in visual juxtaposition, and so aid the detailed study of reservoir features from line to line

(Figure 98). Further, we may ask the computer to strip off everything above the interpreter's reservoir pick, and so improve the juxtaposition.

Further, we can call for a horizontal cross-section instead of the usual vertical one. In seismic terms, this is a *time slice* through the data volume.

Whereas a contour map is the intersection of one reflection event with successive time planes, a time-slice display is the intersection of one time plane with successive reflection events (or successive blacks and whites of the same event; Figure 99). On any one event, however, the time-slice display has the *shape* of the contour map at that time. The process of

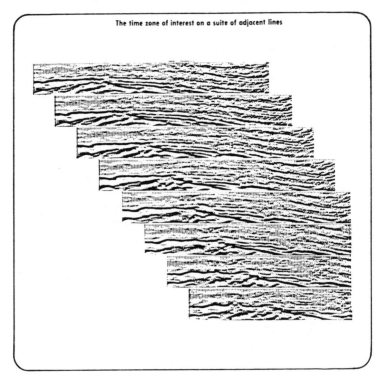

Figure 98 Detailed studies of a reservoir situation on very closely spaced lines. (From Hautefeuille, Cotton 1979; courtesy Oil and Gas Journal)

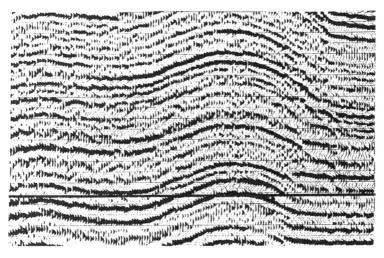

Figure 99 The concept of a time slice (in this case, through a 2-D section).

contouring therefore becomes the tracing of the position of the followed event on successive time slices.

Figure 100 illustrates the construction of a contour map from successive time slices. First, from the time slice at 2632 ms, we trace the outline of some suitable marker (say the outside of the inner bullseye). This is the contour for 2632 ms. Then we trace the position of the same marker on the time slice for 2636 ms; this is the contour for that time. Then we continue the operation on successive time slices, and so build up the contour map (in this case, at a contour interval of 4 ms).

Figure 101 shows how convincingly a time-slice display can indicate the presence and location of faults intersecting a reservoir. With such evidence, it would be foolish to drill delineation wells in random directions.

The time-slice display also has great merit in the study of stratigraphic reservoirs. As we have agreed in earlier discussions, the essential step in recognizing a stratigraphic reservoir seismically is in displaying the seismic indications (however subtle they may be) *in plan*. It is the plan view (in conjunction with the general seismic-stratigraphic environment) which tells us the nature of the reservoir body. If some

Time slices 4 ms apart

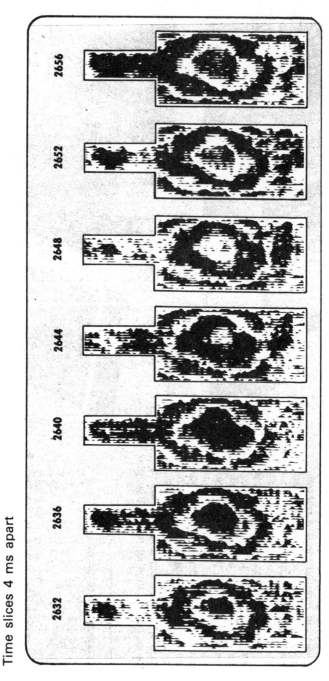

Figure 100 The construction of a contour map from a succession of time slices. (From Brown; courtesy Geophysical Service Inc. and Oil and Gas Journal 1979)

Contour drawn for each level

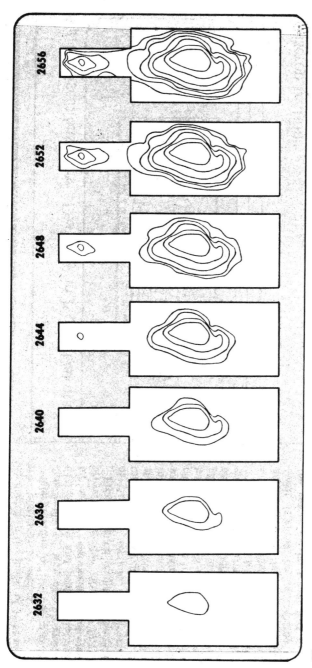

Figure 100 (continued)

135

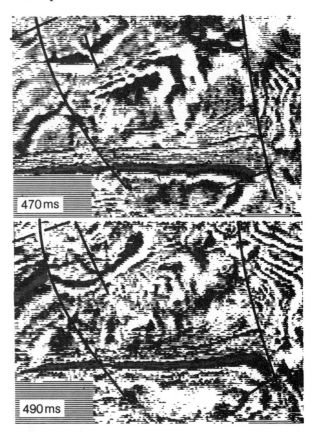

470 ms

490 ms

Figure 101 The appearance of faults on a series of time slices. (Courtesy Prakla-Seismos Report)

indication of the thickness of the anomaly can be preserved, so much the better. And, as we also agreed, it may be important to this recognition to remove the effect of post-depositional structure.

If, then, we pick the structure underlying or overlying the stratigraphic anomaly (which can be done from the vertical cross-sections or from the time slices as above), and then we tell the computer to remove that structure before displaying a *flattened time slice,* we have what we need—an ideal display of the stratigraphic anomaly in plan. A very beautiful illustra-

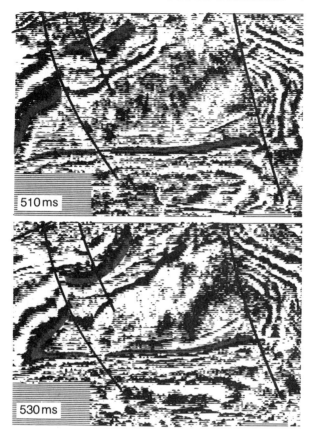

Figure 101 (continued)

tion of this concept (although one which fortuitously did not need supplementary flattening) is given in Figures 102-106. Figure 102 is part of the vertical cross-section for one line of a marine 3-D survey; clearly the message is one of deposition in shallow-water and non-marine environments, but no one would relish the prospect of having to make a detailed interpretation of such a section. Figures 104-106 represent the time slice at the time indicated on Figure 103; they show with superb clarity the intricate meanderings of a river channel. If we were to hit a discovery in one of the point bars of that

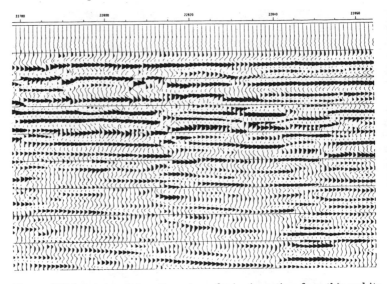

Figure 102 A-not-very-interesting piece of seismic section; from this and its counterparts on closely spaced parallel lines is derived the dramatic time slice of Figures 104-106. (Courtesy Geophysical Service Inc.)

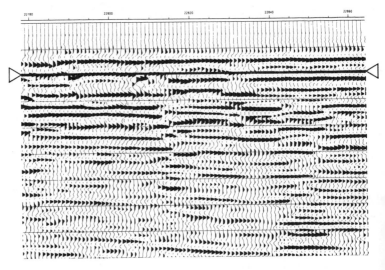

Figure 103 The location of the time slice. (Courtesy Geophysical Service Inc.)

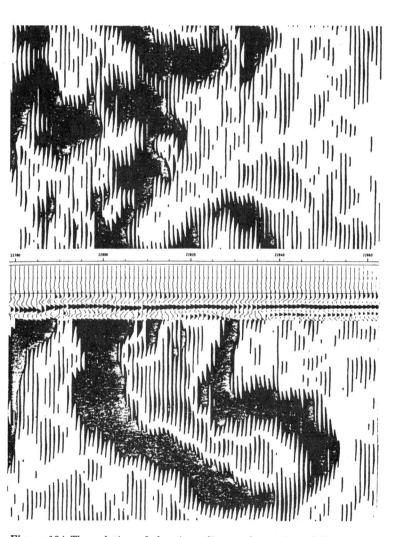

Figure 104 The relation of the time slice to the section of Figure 103. (Courtesy Geophysical Service Inc.)

Figure 105 One of the most beautiful illustrations ever produced by seismic technology; a time slice showing a meandering stream channel. (Courtesy Geophysical Service Inc.)

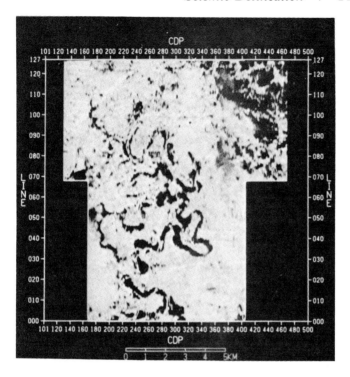

Figure 106 The large-scale picture of the meander belt. (Courtesy Geophysical Service Inc.)

channel, we would certainly know where to drill next. Indeed, the usual approach through delineation wells would scarcely be necessary; we would know quite precisely the clay-sealed edge of the reservoir, and we would have to test only the inboard edge. Further, we could know where to find several more point-bar reservoirs.

Perhaps the most intriguing feature of this figure is that the interpretation leaps out at us *directly from the data*. The display is just as it comes from the computer—untouched by human mind. That way, even the drilling department might believe it.

As a contrast in complexity, Figure 107 illustrates the seismic contour maps obtained from a conventional 2-D survey and a modern 3-D survey. The 2-D survey used a line spacing

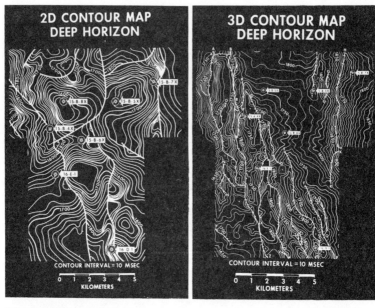

Figure 107 An oft-occurring situation in seismic interpretation: the picture obtained with inadequate data is too simple (and often too optimistic). (From Dahm, Graebner; courtesy SEG)

of 2 km; the 3-D survey a line spacing of 100 m. The comparison reminds us of the difference we so often see between the initial seismic prospect map and the final understanding of the drilled-up field; the initial estimates of structural closure are too optimistic, and the degree and complexity of the faulting is quite underestimated. Since faults are the name of the game, in this locality, the 3-D survey is the only way to establish reliable drilling locations.

Of course, 3-D surveys are more expensive than conventional surveys. Technically, however, the message is clear; for anything except the simplest of structural reservoirs, the 3-D survey is the answer.

If we are conducting a 3-D survey around a discovery well, with the object of mapping the reservoir and minimizing delineation wells, we should certainly conduct a VSP survey as well. If the commercial pressures allow, we should seriously consider keeping the borehole geophone in the hole, at a depth suited to the target reflection, during the recording of at least part of the 3-D survey. That way we really go first-class: a 3-D surface survey, giving us the reservoir configuration far removed from the well, and a 3-D VSP survey, giving us better-resolved details of the reservoir closer to the well.

Working with the Russian VSP experience, the manufacturers of borehole geophones are currently working to provide three or more geophones, on the same cable, at different depths in the hole. Besides speeding the recording of a standard VSP, this arrangement will offer a possibility which would probably be too expensive to contemplate otherwise—that of recording common-depth-point gathers (Chapter 3) in a 3-D VSP. As suggested in Figure 108, an equivalent of the surface-to-surface gather at the left can be obtained with three borehole-geophone positions, as at the right. This not only improves the quality of the reservoir reflections, but also allows determination of the variation of the reflection coefficient with reflection angle; we shall see later in the chapter that this may have new value.

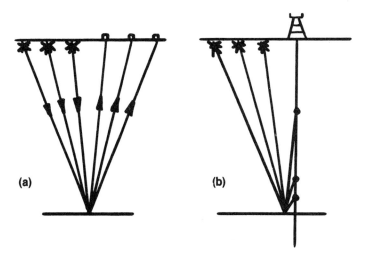

(a)

(b)

Figure 108 Common-depth-point geometry in surface seismics and in borehole seismics.

7.2 Improving Seismic Resolution

Geophysicists have talked about improving resolution—seeing thinner layers—for decades. Then, having talked about it, and feeling better, they went on as before. The reason was that while the chief value of seismics lay in identifying *structure,* resolution really did not matter too much. If the whole sequence went up and down comfortably, the structure could be found anyway; it was only for fault and unconformity traps that some inadequacy was felt.

In the past, therefore, geophysicists were prepared to accept the earth's quirks philosophically. They knew that the earth does not like to transmit short sharp clicks, and insists on turning everything into a thud. So they resigned themselves to using thuds; they designed their field techniques and their equipment to be compatible with thuds, and then optimized the cost of the operation within this limitation. All of which was very reasonable at the time. Now, however, it means that if we are to change our tack we have two things to do: we have to fight the earth, but we also have to change those features of our standard techniques which were previously optimized for structural studies.

And change we must. Nowadays, as we have said, seismics is increasingly expected to do three things: to reveal structure, to assist the identification of the stratigraphy, and to allow measurement of the properties of the rocks. For the second and third of these objectives, and for reservoir studies generally, we feel an acute need for better seismic resolution. We *must* do something. We must see what improvements our technology allows, assess how much improvement can be made for how much cost, and find a new balance between cost and value for the new applications of seismics.

Perhaps some of us sport bumper stickers saying "I'd rather be sailing." For those there is an analogy—the fog horn. The fog horn emits a low-pitched sound, for the very good reason that only low-pitched sounds propagate in fog; high-pitched sounds are quickly absorbed. For years sailors have been thankful for fog horns, and thankful that it was practical to generate sounds loud enough to be heard at great distance in the fog. Only in a relaxed mood, over a tankard in the Jolly Roger, might a sailor observe that the snag with fog horns is that you cannot tell the direction of the sound. "Of course," says the physicist at the next tankard; "That's because the ear cannot tell where low-pitched sounds are coming from. What you want is a fog horn with some high notes too. And then, because the high notes suffer a lot of absorption in the fog, you must emit a lot more power at the high notes—the whole thing will be much more expensive. Will you pay?" "Yes," says the sailor (at last hearing something he understands), "have one on me."

In Chapter 1 we also used the hi-fi analogy, and it may help to use it again. The problem of seismic resolution shares some features with the problem of recording and reproducing music. We are all comfortable with the fact that good reproduction of music requires a system that passes the low notes, the middle notes, and the high notes evenly. If we pass only the low notes we do not hear the piccolo; further, although we may hear the bass we do not hear the actual pluck of the string. If we pass only the middle notes (as when we hear music over the telephone), we cannot resolve the instruments. If we pass only the high notes, everything sounds tinny. And (contrary perhaps to our first expectations) the proper reproduction of the clash of cymbals requires good passage of both high notes and low notes—in fact, all notes.

Let us suppose that a friend is playing radio music to us over the telephone. It sounds thin. We record it, with our recording mike near the earpiece of the telephone. When we play it back, we can improve the sound somewhat by turning up the treble and the bass; we are compensating to some extent the treble-cut and bass-cut effect of the telephone system.

This is what geophysicists call *deconvolution*. Well, more or less.

But suppose that, while we are recording the music, someone starts to run a bath. The mike can hear the music over the sound of the bath, but only just. If now we turn up the treble and bass on playback, the result sounds worse; we know that we must still be getting some improvement of the music, but the benefit is more than offset by the increase in the treble and bass components of the noise of the bath water.

This is deconvolution in the presence of a high noise level. It does not work.

One approach is to ask our friend, who is playing the radio at the other end of the telephone, to turn up the volume at his end. Fine—until the telephone starts to distort.

Another approach is to ask our friend to turn up the treble and bass at his end. The sound goes into the telephone pre-boosted in the treble and the bass, in anticipation of the treble-cut and bass-cut effect of the telephone system. Now, with a little bit of luck, we do not need our own treble and bass boost; the music comes out of the earpiece well balanced, and sounds better. And, although the bath water is still noisy, the music is better than it was (for the same volume level) when we tried to do the compensation at our end. Although this analogy is not exact, it does serve to highlight several features of the seismic counterpart.

When we detonate our explosion, it is a very sharp crack, sharper even than the clash of cymbals. It contains high notes and middle notes and low notes. In the physical processes near the explosion, it loses its low notes and some of its high notes. As the disturbance propagates outwards and downwards in the earth, and as it is reflected back to the surface, it loses more and more of the high notes. What is left, for us to detect with our geophones at the surface, is only the middle notes—as if it had been through the telephone system. To this is added a few crackles (just like the telephone system) and a background of ordinary surface noise (just like the bath water).

The seismic resolution of thin layers, then, is the same problem as the restoration of the full majesty of the clashing cymbals.

If there is no noise—no bath water—then the solution is deconvolution; we turn up the treble and bass after recording. This is excellent, because we can do it all in the processing; the cost is small, and the compensation, being after the event, requires no change in the field procedure. However, in the real world there is always some noise, and this imposes a practical limit on what can be done. Obviously, the practical limit is more serious for a deep reflection than for a shallow one; the deep reflection has lost more of its treble, and is weaker relative to the noise.

This, then, represents the approach and the ability of the established seismic method—the method which must be improved if we are to resolve thinner reservoirs.

How to do it?

One thing is certain: before we do anything else, we must change our recording system so that it can accommodate whatever benefits we are able to realize. We must remove the compromises we have made in the seismic instruments; this is easy and inexpensive, and costs us little more than tape and computer time. More of a change is required in our field techniques; we must sacrifice all those operational conveniences premised on the middle notes only. In practice, this means that the groupings of geophones which we lay out on the surface must be different; we must accept the extra expense of what the geophysicist calls time-and-space-variant arrays. However, the extra cost is mostly in the processing, and is not excessive; we can live with it.

With that done, and with our equipment ready to receive the high notes and the low notes, what do we do next?

One approach is the brute-force-and-ignorance approach. In this we increase the charge of explosive (or the number of charges of explosive) in an attempt to ensure that the music— even the high notes and the low notes—is always louder than the bathwater. To the extent we can do this, we can use deconvolution to balance low, middle and high. The method does not work too well, because large charges of explosive tend to be deficient in high notes. What is more, some of the noise obscuring our reflections is itself generated by the explosive— following paths to the geophone other than the desired reflected paths. In fact, increasing the charge can sometimes increase the noise more than it increases the reflections.

The next approach is to use a large number of small charges. The small charges (¼ lb or 100 g) generate pulses with a good balance of middle and high notes, which is a good start. Then the large number increases the total energy to the point where the music is again louder than the bathwater noise— the point where deconvolution can add the final trim. An increasing amount of work is being done today with this approach. It is tedious and rather costly, operationally, but it produces significant improvements in resolution.

Neither of these methods, however, attacks the problem of the lack of low notes (let us start calling them *frequencies*) generated by an explosive charge, or the problem of the progressive absorption of the high frequencies along the reflection path.

The most promising approach to these problems is given by the Vibroseis system. Alone among seismic sources, Vibroseis allows us to pre-compensate the music before it goes through the telephone.

Suppose that we are at a cocktail party just before it starts to break up. The noise level has been mounting steadily for some time, and by now everyone is having to shout to be heard. That makes the noise worse, so that everyone has to shout louder—which is why the party is about to break up.

If someone on the other side of the room says "Bang!" we shall not hear it above the hubbub. If my name is Al, and the same someone says that, I still will not hear it. But if my name is Arbuthnot Fauntleroy Featherstonehaugh, and if the same someone says that, I shall swing round to see who is calling. The length of the "signal" allows the brain to recognize it against a high level of noise, even though the noise may drown a few syllables of the signal.

The principle of Vibroseis, then, is to emit *long signals of highly characteristic form,* and to recognize these signals when they are reflected back. Although the power level at which we emit them should obviously be as high as practicable, this power level does not have to be as high as that of the traditional seismic explosion; this is because of the increased recognition ability associated with the long signal. The recognition process (we call it *correlation*) takes in "Arbuthnot Fauntleroy Featherstonehaugh," and puts out "Me!"

In usual practice, the long signal of characteristic form is a *sweep;* it starts as a steady low note, and the pitch rises to a high note.

In our musical analogy, we said that to reproduce the clash of cymbals we must be able to pass the low notes, the middle notes and the high notes evenly; the clash (though our ears may not tell us so) contains all these notes. Similarly the seismic bang contains a range of notes—all at once, so to speak. The Vibroseis sweep contains the same range of notes, but spread out in time—first the low frequencies, then the middle frequencies, and then the high frequencies. The recognition or correlation process puts them all together again, so that the result is a bang, visually indistinguishable from the traditional one. And though the reflected sweep may be below the noise level, the correlated bang is above it.

Now we can see how the Vibroseis system offers a distinctive potential for improving seismic resolution. We can spend *more time* sweeping the frequencies which are going to be reduced along the seismic path. We pre-compensate the emitted signal by making it richer in high frequencies and low frequencies.

Loosely, geophysicists talk about the low frequencies and the high frequencies, which we need for improved resolution, as being "lost" along the seismic path. They are not lost; they are attenuated. If we send out enough extra low frequencies and enough extra high frequencies to overcome this attenuation, we can hope to achieve a final signal with an even balance of low, medium and high frequencies.

Of course, at very low frequencies and at very high frequencies this attenuation is enormous. We would have to spend so long sweeping these frequencies that we would be on each shot-point for a month. But the point is established: there is no technical limit to seismic resolution with the Vibroseis system, and we can resolve layers a metre thick if someone will pay for it.

Any geophysicists reading this will be fretting at all the talk of Featherstonehaugh and his cymbals. So, just for them, we should take a moment to set down a specific Vibroseis example. Excuse us, while we talk shop for a paragraph or two.

We consider the case of Figure 109, in which the smooth curve represents the loss of signal-to-noise ratio at the low and high frequencies, relative to a peak at about 25-30 Hz.*

We stress that the vertical axis is signal-to-noise ratio, not signal spectrum; the signal spectrum can be extended in the

---

*The hertz, Hz, is the unit of frequency—the number of vibrations per second.

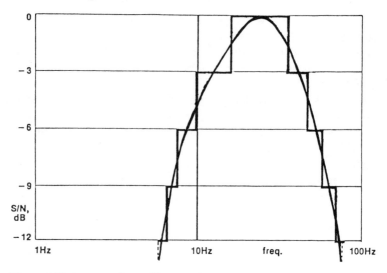

Figure 109 An example to illustrate how improved resolution is obtained with the Vibroseis system, by spending more time at the low and high frequencies.

processing (by deconvolution) but the signal-to-noise ratio must be extended in the field. The signal-to-noise ratio (which we will assume is adequate, as in the usual case, at the middle frequencies) is down to one-half (− 6 dB) at 8.9 Hz and 50 Hz, and down to one-quarter (− 12 dB) at 5.9 Hz and 70 Hz; this is after the best that can be done by hard-driving the vibrator and by other signal-to-noise enhancement techniques. Then, as suggested by the ziggerat line, a step-wise compensation of this response can be made by dividing the sweep into 9 subsweeps of the indicated frequency ranges, and by emitting multiple subsweeps at the higher and lower frequencies for each one necessary at the middle frequencies. If we do the arithmetic, we find that the standard practice of sweeping (say) 15-48 Hz is likely to take 2.7 minutes at each shot-point. Extending the frequency band to 5½-80 Hz without compensating the attenuation at low and high frequencies would involve 6.1 minutes (but do little good). Extending the band to 5½-80 Hz and providing full compensation involves 21.6 minutes at each shot-point.

We make no bones about it: resolution is expensive. We can get it, but it is expensive.

Of course, the compensation does not have to be stepwise, as described above; we can obtain a smooth compensation by programming the sweep rate accordingly. Undoubtedly this will become the normal method as appropriate sweep generators become standard equipment. The virtue of the step-wise approach through subsweeps is that it can be done with minor modification of earlier equipment.

But we reiterate: these fancy techniques are for well-defined reservoir problems of limited extent. They are not for general exploration use, until we can find ways to reduce the cost. In fact, the application which is just crying out for these techniques, at present, is reservoir delineation by 3-D VSP. For this application, as we have noted before, the degradation of the signal in the near-surface is halved, all the technical factors are on our side, and the cost-benefit is more easily defined.

7.3 More Rock Properties

The traditional seismic measurement is reflection time. The seismic method found a lot of petroleum by mapping structures, using nothing but reflection time.

Then the seismic method found a lot of gas by studying reflection amplitudes; gas in sandstone reservoirs in a young sand-shale sequence was found to generate bright spots.

Then increased sophistication in the measurement of velocity allowed us to begin putting some faith in the *interval velocity* indicated for a layer. However, the interval velocity is not uniquely interpretable in geologic terms; it depends on the lithology, the porosity, the cement, the saturant(s), the degree of fracturing and the presence of overpressure.

Then increased sophistication in the treatment of reflection amplitudes allowed us to begin putting some faith in measurements of rock *hardness* (properly, acoustic impedance). And, as set out in Chapter 3, the combination of a measure of velocity and a measure of hardness gives us a measure of *density*. Density, we know, also depends on the lithology, the porosity, the cement, the saturant(s) and over-pressure, but it is less dependent than velocity on the presence of fracturing.

Density and velocity are very important rock properties, particularly because of their dependence on porosity. If we could compute from surface-seismic data the density and velocity of every reservoir, we would be happy indeed. The problem, of course, is resolution; to do that we need to have well-separated reflections from top and bottom of the reservoir—which is easy if the reservoir is 100 m thick but less so if it is 10 m thick. However, as we have just concluded in the last section, there is no barrier of principle; provided the reservoir does generate usable reflections we can tackle the 10-m problem if the answer is worth the cost.

It is natural to ask if there are other measurements which might be made on seismic reflections, and which might reduce the lithologic ambivalence of the velocity and density measurements.

One proposal is absorption. It seems to be established that a sand (particulary an angular sand) absorbs high frequencies from the seismic signal more than does a clay. The absorption is further affected, in a complicated way, by the proportions of gas and liquid in the pores. However, no significant commercial use of absorption measurements is current. As we improve our resolution (by the techniques of the last section), the potential for absorption measurements will increase.

Another proposal is shear-wave measurements. Traditional seismic techniques use compressional or P waves—the kind of bang generated by an isolated explosion, in which the rock particles are *P*ushed outwards. Shear-wave techniques use shear or S waves, in which the rock particles are *S*haken from side to side. Shear waves can be generated by various combinations of explosives and trenches, or by giant sideways-striking hammers, or by sideways-vibrating vibrators; they are received by sideways geophones, planted by sideways people.

The hopes for shear-wave seismics are three. First, shear velocities are about half of compressional velocities, so that *if* the same frequencies can be used the resolution is immediately improved by a factor of two; unfortunately the frequency range usually drops also, so this benefit is small. Second, shear waves are less affected by the saturant than are compressional waves, so that a comparison of shear and compressional results may highlight a change of saturant. Third, the ratio of shear velocity to compressional velocity is more indicative of the lithology than the compressional velocity alone.

Current use of shear-wave techniques in the West is limited to a few major companies. The problems, however, are being worked out. One of the problems is the simple identification of the "same" reflection on the two sets of results; for this a compressional VSP and a shear VSP are invaluable.

In fact, the best place to practice all these newly emergent techniques is in a discovery well—in reservoir studies using 3-D VSP and surface methods. The origin of all reflections can be authoritatively identified, the velocities are known precisely, and the logs are available to aid the identification. The correlation between the nature of the reservoir and its seismic response can be established with certainty at the borehole, so that then we may concentrate on the mapping and understanding of the variations in this response away from the borehole.

An interesting transition between compressional and shear approaches is given if we keep the borehole geophone at one depth and progressively increase the distance of the seismic source from the wellhead. The usual narrow-angle reflection gives way to a wide-angle reflection, which gives way to refraction. Although this requires the additional complexity of sideways geophones in the borehole tool, and although the interpretation problems are formidable, there can be no doubt that additional valuable information is waiting for us here. Further, as we discussed in connection with common-depth-point VSP work, the use of several geophone positions in the hole allows us to make studies of the effect of changing the reflection angle without changing the reflection point.

Roll on the day. It is in borehole seismics that the next Great Leap Forward will occur.

Finally we should discuss the seismic log (or impedance

log, or pseudo-log, or Seis-log, or any of a host of other proprietary names). This is not an additional seismic measurement, not an additional rock property; it is an alternative display for traditional results. However, since it facilitates comparison of seismic and well-log data, new geological insights into the nature and meaning of the seismic reflectors may be obtained.

For present purposes, we may think of the pseudo-log as obtained by integration; Figure 110 reproduces the essence of our discussion in Chapter 3. This is the *pseudo-log-without-trend;* it oscillates both sides of a zero line, just as a seismic trace does.

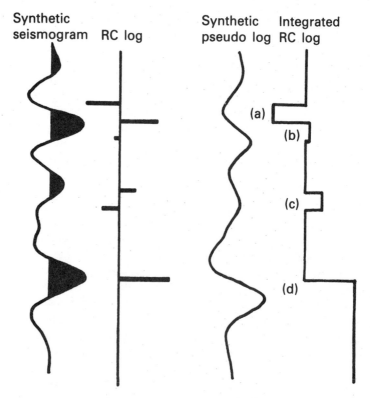

Figure 110 Reprise of Figure 36.

The pseudo-log, we remember, attempts to show us layers rather than contacts. To a first approximation, a deflection to the right represents a harder layer, and a deflection to the left a softer layer. The snag, we also remember, lies in the additional spurious deflections introduced by the oscillatory nature of the seismic pulse.

To the extent that the variations of hardness indicated by a pseudo-log may be thought of as variations of velocity, we can make a *pseudo-log-with-trend* by superimposing the pseudo-log on the interval-velocity measurement obtained from seismic velocity analysis. We are combining, in effect, the message of hardness from reflection *amplitude* with the message of velocity from reflection *moveout*. Then, as a final refinement (and to the extent that we are confident in doing so) we may remove that part of the interval-velocity variation which is due to compaction alone; this probably brings us as close as we can be, today, to a cross-sectional display of lithology.

The latter two types of pseudo-log involve a considerable range of values, and do not look well in conventional cross-sectional form. The usual solution is to contour the pseudo-log traces, and to colour the contour intervals with suitably contrasting colours. Although examples of this display cannot be reproduced here, illustrations in full colour abound in the advertising pages of *Geophysics* and *Geophysical Prospecting*.

8

The Proper Functions
and the
Cost-Effectiveness
of Seismic Tools

Years ago, lines of demarcation were drawn between the concerns of the exploration geologist, the geophysicist, the production geologist and the engineer.

The line between the exploration geologist and the geophysicist has almost disappeared. Geophysics is the eyes and ears of the exploration geologist, and he would be foolish to work without it. On the other side, modern seismic-stratigraphic interpretation is impossible without liberal help from geology.

Slower to erode is the line between the geophysicist and the reservoir experts. But it must yield. Geophysics is the means by which the production geologist and the engineer can look laterally into their reservoir, outwards from the borehole. On the other side, seismic interpreters must insert a better understanding of the nature of real reservoirs into their interpretations.

Not only do times change, and understandings change,

and perceptions of our functions change; technology changes—perhaps at different rates in different disciplines—and costs change.

All of this highlights the conclusion that the old judgements we used to make may no longer be sound. The old question "Do we shoot more seismics, or drill another hole?" may have a different answer today, as drilling costs rise and seismic techniques improve. Further, the correct answers to many everyday questions may be changing: Do we need a sonic on this hole? Why should we shoot it for velocity? How can we tell whether there are two reservoirs without drilling? And isn't it true that nothing finds oil and gas but the bit?

So we should review the present status and present costs of different ways of studying a reservoir.

First, we should set out the *sample volumes* which contribute to seismic measurements.

The sonic log measures the vertical seismic velocity, at kilohertz frequencies, of a very small sample of rock close to the borehole. Further, it usually measures the velocity of the fastest path only; the local fastest path may not be representative of the material in bulk. The long-spaced sonic log samples more rock, but, like all logs, still tells us next to nothing about the vast bulk of the reservoir away from the hole. Against that, we must also add that, in respect of fluid flow, the zone near the hole is the most important.

The check-shoot and the VSP measure the vertical bulk velocity, at seismic frequencies, of a sausage-shaped sample of rock between the surface source and the geophone.

Velocity analysis from surface seismics measures a quantity which, for present purposes, we will accept as the horizontal bulk velocity, at seismic frequencies, of a large triangle of rock representing the geometry of the common-depth-point gather (Figure 8). In fact, it is larger than this, since each path from source to reflector to receiver itself involves a sausage-shaped sample of rock along that path. Typically, the total sample might be a trapezoid having a horizontal extent of 2500 m (1½ miles) on the surface, and 300 m (1000 ft) on the reflector.

Interval-velocity computations from surface seismics measure a quantity which, for present purposes, we will accept as the horizontal bulk velocity, at seismic frequencies, of a disc-shaped volume whose upper and lower surfaces more-or-less represent the intended interval and whose diameter is typically 300 m (1000 ft).

The distinction between horizontal and vertical velocities may be material, because the velocity of many rocks (particularly shales with marked stratification) is significantly greater parallel to the bedding than perpendicular to the bedding. The distinction of frequency may be material because the velocity at sonic-log frequencies tends to be slightly higher than that at seismic frequencies.

The *area of the reflector* sampled by a reflection depends on several variables, of which the most important are the source and the length of the reflection path. This is the reason for our previous observation that the horizontal resolution of a 3-D VSP is better than that of surface seismics; the path length is less, and because the downhole noise is less than the surface noise the source can be smaller (and hence sharper). Usually, the reflector area sampled is a circle whose diameter is in the range 200-750 m (650-2500 ft) for a surface-to-surface reflection, and in the range 150-250 m (500-800 ft) for VSP reflection.

If a top-reservoir reflector is flat over the area of this circle, and if the reservoir and cap-rock properties are constant over the same area, the top-reservoir reflection. amplitude truly represents the contrast of hardness between cap-rock and reservoir. If the properties vary, the reflection amplitude represents the average. If the reflector is not flat, or the contrast of hardness is transitional, the reflection amplitude changes (usually, decreases). In particular, if the top-reservoir contact is cut by one or more faults within the area, the reflection amplitude is decreased.

The point of all this, obviously, is that there are limits on the size of feature which we can see with the seismic method, and more onerous limits on the size of feature which allows us to do calculations on it. We cannot see from the surface a feature which is only a few metres in horizontal extent; a few metres thick maybe, but not a few metres wide. And to do reliable calculations of the properties of the feature, it had better be several hundred metres wide.

In one sense, this is not all bad. The seismic indications of reservoir properties are very gross, very smeared; but they may be more indicative of the reservoir as a whole than are the logs run in a well.

So let us accept that the borehole logs (and the depositional geology inferred from them) represent a sample of the earth different from that observed by seismics, and that this must be accommodated when we consider them together.

With that said, let us stress again the ideal technical approach to the delineation of a reservoir. The order given is not necessarily the actual order.

- We run our suite of logs and borehole measurements (including sonic and density) in the discovery well.

- We make our first judgements of the nature of the reservoir, using the cores and/or cuttings, the drilling information, the logs and the borehole measurements, the tests, and the pre-existing seismic data used in both a structural and seismic-stratigraphic way.

- In particular we renew the search for a fluid-contact reflection on the pre-existing seismic data. To aid this, we may run an initial synthetic seismogram (without benefit of calibration) over an interval including the reservoir and some suitable marker reflection. If we find a fluid-contact reflection, we draw initial conclusions about the extent and division of the reservoir.

- We run a VSP. We use this:
  - to alert us of intrusions or major faults close to the borehole,
  - to see below the drill,
  - to calibrate the sonic,
  - to understand the reservoir reflections, and to measure their reflection coefficients,
  - to generate the expected seismic record at the surface, by projection of the upcoming waves to the surface,
  - to decide whether the reservoir-delineation problem is soluble by seismics, and what field techniques are appropriate if it is.

- We prepare the definitive synthetic from the calibrated sonic, using all our wits (and the VSP surface-projection) to get a reliable match with the pre-existing seismic data through the well.

- We perturb the input to the synthetic to represent a range of likely changes in the thickness and/or porosity of the reservoir, we construct the corresponding range of synthetics, and we search the pre-existing seismic data for matches away from the well.

- Somewhere in this sequence we make a first judgement whether this will be a one-well field (so that our problem is merely to estimate reserves), or whether we also require to locate further wells. The following steps assume the latter.

- We make the judgement whether to mount a full 3-D surface survey, or whether to run a few surface lines through the well, or whether to use a 3-D VSP. If the reservoir is not expected to extend from the well more than one-third of its depth, we probably prefer the 3-D VSP on resolution grounds (though we accept that the interpretation may be tough).

- If the decision is a full 3-D surface survey, we adopt field and processing techniques shown to be appropriate by the previous synthetic and VSP work. Further, we remember our discussions of side-swipe and 3-D migration, and are careful to cover sufficient area beyond the expected limits of the reservoir. Then we make our interpretation, using both vertical cross-sections and horizontal time slices. In particular, we attempt to remove post-depositional structure from the time slices, hoping that the sedimentary nature of the reservoir will be apparent from its shape in plan. We search also for evidences of reservoir bodies which may overlap but be separated, such as bars and distributary channels, or a succession of regressive sand pulses. Also, of course, we map the faults, and the fluid contact if it is visible.

- If the decision is for a 3-D VSP survey, on grounds of resolution or cost or other factors, we first use all our information to be sure that we locate the geophone(s) optimally. We decide whether the usual six-star arrangement of surface sources is sufficient, and extend the source lines to at least double the expected limits of the reservoir.

- Finally, whatever field method we use, we seek to establish the area and closure of the reservoir body, the position of fluid contacts, the division of the reservoir by faults, and the likelihood of division of the reservoir by shale breaks. Further, if the resolution we achieve is appropriate to the thickness of the reservoir and its encasing beds, we seek by study of reflection amplitudes and velocities (calibrated at the well) to extend our knowledge of reservoir porosity away from the well, thus we can hope to indicate zones where the porosity is better than at the well, and zones which represent porosity barriers.

We have stressed that this is what we would like to do technically, in the absence of economic constraints. But in the real world, we must decide what course represents the best compromise between cost, business prudence and politics, and information.

Obviously, the optimum compromise depends so much on the circumstances that we cannot give pat answers. But we can identify some of the relevant circumstances, and look at current costs.

One of the most important variables is the depth of the reservoir. Drilling costs increase steeply with depth. Logging and VSP costs increase more-or-less linearly with depth. Surface-seismic costs actually decrease with depth, provided that the resolution obtained is sufficient; if not, the high cost of increasing resolution increases steeply with depth. The area of reservoir which may be covered by a 3-D VSP increases with depth. All of this means that we are unlikely to be calling for much geophysics if we have a simple reservoir at shallow depth, where a step-out well can be drilled for $50,000.

Another important variable is whether the well is

offshore. Logging and VSP costs are not much changed. 3-D VSP costs are not much more, and surface-seismic costs may be significantly less; however, there are currently no possibilities of improving resolution by anything other than the brute-force-and-ignorance approach. Drilling costs, of course, are much increased.

So the chances are that we shall be doing a great deal of geophysics if our reservoir is deep and the well is far enough offshore to need a big semi-submersible or a drillship. We would be crazy to spend $20 million on a delineation well without first spending a few percent of that on geophysics.

Onshore, of course, the issue is complicated by problems of leases and access and permits. Both geophysical and well costs are affected by mobilization expenses, but not necessarily in the same proportion. Where drilling cannot be commissioned on a turn-key basis, unexpected expenses may result from mechanical problems in the hole; surface-seismic work is usually turn-key. In some situations, VSP and 3-D VSP work involve loss of rig time, while 3-D surface seismics may require shutting down the rig for a period; however, the natural time to do these surveys is during the initial production period, when the only loss is the deferral of some production.

Obviously there are a thousand factors which affect the final costs, but it may still be worth setting down some broad ranges:

Drilling costs (dry-hole): $50,000—30 million

Extra cost to run sonic log: $ (few thousand)

Initial synthetic: $ (few hundred)

VSP and sonic calibration: $5,000—10,000*

Synthetic (with perturbed $1,000
    variations):

---

*Plus rig time or shut-down time if rig is on location.

3-D VSP:	$10,000—25,000*
3-D VSP with improved resolution:	$30,000 upwards*
Surface seismics, six-star:	$20,000—30,000*

3-D surface seismics

Onshore, 50 km²:	$200,000*
Offshore, 100 km²:	$1,000,000

These figures illustrate that a great deal of scope exists for geophysics if we can guarantee that the geophysics will save at least one wasted well (and the delay in petroleum production that is represents).

There are no guarantees, of course. Further, we must acknowledge that even a wasted well gains some valuable positive information. We must acknowledge that some reservoirs (particularly, very porous oil-saturated sands in shale) do not generate usable seismic reflections. And that many reservoirs are too thin for the seismic approach to be cost-effective. Nevertheless, despite all this, we place on the record here, fairly and squarely, that: *We should be using geophysics for the location of delineation and development wells far more than we presently do.*

---

*Plus rig time or shut-down time if rig is on location.

Bibliography

Anstey, N. A., 1977, Seismic interpretation: the physical aspects: Boston, IHRDC.

———— 1980, Seismic exploration for sandstone reservoirs: Boston, IHRDC.

———— 1981, Seismic prospecting instruments, volume 1 (2nd edition): Berlin, Gebruder Borntraeger.

Brown, A. R., 1979, 3-D seismic survey gives better data: Oil and Gas J., v. 77, n. 45, p. 57–71.

Bruce, C. H., 1973, Pressured shale and related sediment deformation: mechanism for development of regional contemporaneous faults, AAPG Bull., v. 57, n. 5, p. 878–886.

Bubb, J. N., and Hatlelid, W. G., 1977, Seismic recognition of carbonate buildups: Tulsa, AAPG Mem. 26, p. 185–204.

Coffeen, J. A., 1978, Seismic exploration fundamentals: Tulsa, Petroleum Publishing Company.

Dahm, C. G., and Graebner, R. J., 1982, Field development with three-dimensional seismic methods in the Gulf of Thailand—a case history: Geophysics, v. 47, n. 2, p. 161.

Dix, C. H., 1981, Seismic prospecting for oil: Boston, IHRDC.

Dobrin, M. B., 1976, Introduction to geophysical prospecting: New York, McGraw-Hill, Inc.

Hautefeuille, A., and Cotton, W. R., 1979, Three-dimensional seismic surveying aids exploration in the North Sea: Oil and Gas J., v. 77, n. 45, p. 79.

Lindsey, J. P., Dedman, E. V., and Ausburn, B. U., 1978, New seismic techniques define strat, lith details: World Oil, v. 186, n. 6.

McQuillin, R. M., Bacon, M., and Barclay, W., 1979, An introduction to seismic interpretation: Houston, Gulf Publishing Company.

Mitchum, Jr., R. M., Vail, P. R., and Sangree, J. B., 1977, Stratigraphic interpretation of seismic reflection patterns in depositional sequences: Tulsa, AAPG Mem. 26, p. 117–133.

Mitchum, Jr., R. M., Vail, P. R., and Thompson, III, S., 1977, The depositional sequence as a basic unit for stratigraphic analysis: Tulsa, AAPG Mem. 26, p. 53–62.

Sangree, J. B., and Widmier, J. M., 1977, Seismic interpretation of clastic depositional facies: Tulsa, AAPG Mem. 26, p. 165–184.

Schramm, Jr., M. W., Dedman, E. V., and Lindsey, J. P., 1977, Practical stratigraphic modeling and interpretation: Tulsa, AAPG Mem. 26, p. 477–502.

Sheriff, R. E., 1973, Encyclopedic dictionary of exploration geophysics: Tulsa, SEG.

—— 1978, A first course in geophysical exploration and interpretation: Boston, IHRDC.

—— 1980, Seismic stratigraphy: Boston, IHRDC.

Vail, P. R., Mitchum, Jr., R. M., and Thompson, III, S., 1977, Relative changes of sea level from coastal onlap: Tulsa, AAPG Mem. 26, p. 63–81.

Waters, K. H., 1978, Reflection seismology: New York, John Wiley & Sons.

Index

166